# SUPERMASSIVE

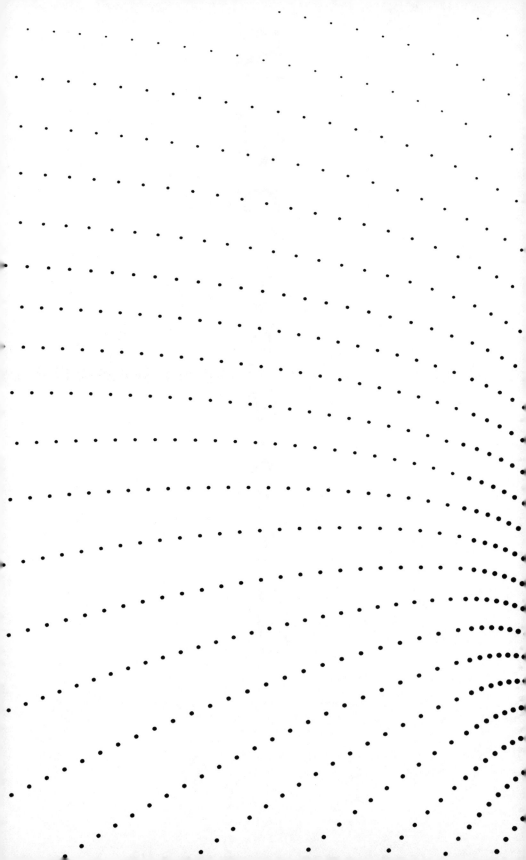

# SUPERMASSIVE

## BLACK HOLES AT THE BEGINNING AND END OF THE UNIVERSE

JAMES TREFIL AND
SHOBITA SATYAPAL

SMITHSONIAN BOOKS
WASHINGTON, DC

© 2025 by James Trefil and Shobita Satyapal

Published by Smithsonian Books
PO Box 37012, MRC 513
Washington, DC 20013
smithsonianbooks.com

Director: Carolyn Gleason
Senior Editor: Jaime Schwender
Edited by Gregory McNamee
Designed by Jody Billert and Faceout Studio/Jeff Miller

This book may be purchased for educational, business, or sales promotional use. For information, please write the Special Markets Department at the address or website above.

Library of Congress Cataloging-in-Publication Data

Names: Trefil, James, 1938- author. | Satyapal, Shobita, 1966- author.
Title: Supermassive : black holes at the beginning and end of the universe / James Trefil and Shobita Satyapal.
Description: Washington, DC : Smithsonian Books, [2025] | Includes index.
Identifiers: LCCN 2024044072 (print) | LCCN 2024044073 (ebook) | ISBN 9781588347893 (hardcover) | ISBN 9781588347909 (ebk)
Subjects: LCSH: Black holes (Astronomy)
Classification: LCC QB843.B55 T738 2025 (print) | LCC QB843.B55 (ebook) | DDC 523.8/875—dc23/eng/20241119
LC record available at https://lccn.loc.gov/2024044072
LC ebook record available at https://lccn.loc.gov/2024044073

Printed in the United States
29 28 27 26 25     1 2 3 4 5

# CONTENTS

# PROLOGUE

———

Out of the cradle
onto dry land
here it is
standing:
atoms with consciousness;
matter with curiosity.

Stands at the sea,
wonders at wondering: I
a universe of atoms
an atom in the universe.

*Richard Feynman,* "What Do You Care What Other People Think?":
Further Adventures of a Curious Character

I t's 7:00 am, and I'm driving down a mountain in a large four-wheel drive vehicle. Barely awake, exhausted, I notice that my fingers have turned blue. I have never driven a truck before, and now I am hurtling down one of the steepest and toughest roads in the world. I feel the brakes overheating, the truck skidding, and I watch the sunrise wondering if I'm about to drive off the edge. I shouldn't be driving this truck, descending from an altitude of 13,796 feet (4,205 m) at the summit of Mauna Kea—the second highest peak on Earth, and by some measures the highest—where I have spent the night observing the night sky. The black sand beaches merge into the deep blue sea below. I look to the right and see that my backpack has fallen off the passenger seat. Inside is my computer, which holds the data we obtained from last night's observing run. There is a little secret of the universe hidden inside that backpack. My fear and sense of a separate self dissolve into the sky above as I reflect on the data on my computer, a gift from a few photons that traveled 47 million light-years to the infrared detector attached to the end of the NASA Infrared Space Telescope (IRTF), where I had been granted four nights of observing with an infrared spectrometer.

Mauna Kea is an inactive volcano on the island of Hawai'i that is home to several large observatories. According to Hawaiian tradition it is a deeply sacred and revered mountain. As I drive down the winding road from the valley below to the summit above, I honor its ancient history and feel gratitude for all that has been enabled by Mauna Kea in my lifetime. Founded in the late 1950s, the observatories at its top have allowed humanity to peer into the universe at wavelengths that were unobservable through most of human history. At sea level there are almost 5.5 quadrillion tons of atmosphere above us, providing the air we breathe, protecting us from harmful radiation from space, creating the pressure we need for the liquid water that flows on the surface below. This indispensable blanket that protects our planet makes infrared astronomy hard. Very hard. It's why I'm on this mountain, barely awake and freezing, gripping the steering wheel. In fact, the science that I want to do is impossible to carry out from the ground at lower altitude. This is because the infrared photons that we have collected, now stored on my computer, cannot make it through the atmosphere. They are absorbed by the precious water molecules found in Earth's atmosphere. For that reason, we have to build observatories at high altitude or launch them into space in order to be able to do infrared astronomy.

Last night we peered into the heart of the galaxy NGC 1068 to study its central black hole. NGC 1068, or Messier 77, is a spiral galaxy almost 47 million light-years away in the constellation Cetus. It was discovered by the French astronomer Pierre Méchain in 1780, before we even knew that galaxies outside our own existed. There, obscured by clouds of gas and dust, invisible at optical wavelengths, lies a feeding black hole. It's not just an ordinary black hole that forms from the death of a massive ordinary star, but a *supermassive* black hole with an estimated mass of over 100 million times the mass of the Sun.

How did it form? How does it grow? What does it do to the galaxy that surrounds it, and what will be its ultimate fate? We don't know the answers to these questions, at least not yet. In the past four nights of observing on this revered mountain we were fortunate to see several galaxies in which we think supermassive black holes reside. The data recorded the light emitted by matter orbiting close to the black hole, swirling around the center at speeds of several hundred kilometers per second. By measuring the speed of gas, we can measure the mass of the central supermassive black hole. We can also study the properties of stars around the black hole. How many are there? How old are they? How are they influenced by the black hole? It turns out that matter isn't just sucked into black holes. It is also often being expelled from the central supermassive black hole at tremendous velocities, sometimes approaching the speed of light. In fact, black holes can fling out the mass equivalent of an entire galaxy deep into the void of intergalactic space.

All of this information about the stars, gas, and the supermassive black holes is encoded in the spectra stored in my laptop in my backpack. I sit, sleep-deprived, in a hard plastic seat in the airport waiting to board a fourteen-hour flight back home. It will take months for us to fully reduce this data, but I am going to get started on the plane.

Join us as we take you through a journey from the early history of black holes, of how they arose as mathematical predictions of Einstein's theory of general relativity to how they became real astrophysical objects that can be almost 100 million times the mass of the Sun. We will bring you up to speed on the latest research on these invisible beasts now found to be lurking in the hearts of galaxies only a few hundred million years after the Big Bang.

My airplane ride will last a bit more than half a day. Getting to this point has required thousands of years of scientific discovery. More recently, it has turned on the critically important discovery that black holes exist in the first place, and that some are more massive than others. This scientific journey takes us to the desk of the brilliant scientist Albert Einstein, whose general theory of relativity helped us predict black holes before we even saw them. Our whirlwind tour of the universe begins just a few hundred million years after the Big Bang, 13 or so billion years ago.

# 1

# FROM NEWTON
# TO EINSTEIN

---

The Khazars believe that deep in the
inky blackness of the Caspian Sea there is
an eyeless fish that marks the only correct
time in the universe.

*Milorad Pavič*, Dictionary of the Khazars

T he universe used to be such a simple place. Events ticked along
like the hands of a clock, driven by simple laws that were not
only knowable, but known. Behind the surface complexity of
the world there lurked an amazing simplicity.

Perhaps nothing better exemplifies this way of looking at the universe than the work of the English scientist Isaac Newton (1642–1727).
In a period that will be familiar to modern readers, the appearance of
the Black Plague in England led to what we would call today a lockdown,
and Newton was forced to spend eighteen months in isolation on his
family farm. During those eighteen months, he developed differential

calculus, integral calculus, the law of universal gravitation, the theory of color, the laws of motion, and a few miscellaneous mathematical theorems. All we can say is that if reading that list makes you wonder what you've accomplished in the last few years, you aren't alone.

For our journey into the heart of black holes, the most important item on Newton's list is the law of universal gravitation. As Newton stated it, this law says that there is an attractive gravitational force between any two objects in the universe. The size of the force depends on the mass of the two objects and the distance between them. Double the mass of one of them, and the gravitational force doubles. Double the mass of both of them, and the force will be four times as strong. The gravitational force gets weaker the farther apart the two objects are: double the distance, and the force will be four times smaller; triple the distance, and it will be nine times smaller; and so on.

The equation that describes the gravitational force between bodies of mass, $m$ and $M$, separated by a distance, $D$, is:

$$F = \frac{GmM}{D^2}$$

where $G$ is a number known as the gravitational constant.

## THE NEWTONIAN UNIVERSE

According to Newton, who described the event later in life, his discovery of universal gravitation was triggered by seeing an apple fall from a tree at the same time he saw the Moon in the sky. He realized that there had to be a force acting on the Moon, otherwise it would fly into space instead of staying in orbit. He also realized that there had to be a force acting on the apple to make it fall. In one of those insights that seem so obvious in retrospect but that takes a genius to see for the first time, he asked whether those two forces—one operating on Earth, the other in the heavens—could be one and the same. With this insight he paved the way for generations of astronomers to study the skies, secure in the knowledge that the phenomena they saw in their laboratories were the same as the ones they saw through their telescopes. Philosophers regard this insight as the first of a series of great unifications that characterize the history of science.

While Newton recognized that the force of gravity existed, he couldn't understand how it worked. In Newton's world, forces could be transmitted only by contact, something like what happens when billiard balls collide. In his theory, the action of gravity was instantaneous. This apparently bothered him, but he necessarily accepted that idea, since it would be years before the first estimates of the speed of light or the speed at which gravitational forces actually move would be available. The discovery that matter is made from atoms, and that atoms interact with each other through electromagnetic force, would also occur well beyond Newton's lifetime. He pondered the nature of gravity in a letter to his friend Richard Bentley in 1692:

> It is inconceivable that inanimate Matter should, without the Mediation of something else, which is not material, operate upon, and affect other matter without mutual Contact. . . . That Gravity should be innate, inherent and essential to Matter, so that one body may act upon another at a distance thro' a Vacuum, without the Mediation of any thing else, by and through which their Action and Force may be conveyed from one to another, is to me so great an Absurdity that I believe no Man who has in philosophical Matters a competent Faculty of thinking can ever fall into it. Gravity must be caused by an Agent acting constantly according to certain laws; but whether this Agent be material or immaterial, I have left to the Consideration of my readers.

What is amazing to us, as we follow Newton's thoughts about the nature of gravity, is how close he came to understanding the way that the gravitational force arises in modern post-Einsteinian theories. In string theory and quantum loop gravity, the two leading contenders for a "theory of everything," the gravitational force is mediated by the exchange of massless particles called gravitons. Whether you want to call these particles, which were undiscovered in Newton's time, "material" or "immaterial agents," to use Newton's language, there does seem to be an uncanny foreshadowing of future events in his musings.

The law of universal gravitation has proved to be a reliable tool for explaining familiar phenomena like the orbits of the planets and the path of space probes. It is only when you get to massive objects like black holes that you need a better theory, in this case the theory of general relativity, about which more soon.

First, though, consider two important aspects of the Newtonian worldview. One of them can be considered the philosophical basis of general relativity, and the other had to be overcome before progress on relativity could be made. In Newtonian mechanics, all observers see the same laws of nature operating regardless of their frame of reference, provided that the frames of reference are moving with constant velocity. We call this the principle of special relativity. The second thing built into the Newtonian worldview is the hidden assumption that there exists a correct universal time underlying all laws of nature.

Suppose you are standing at the side of a road, and a friend drives by in a car. As your friend passes you, she drops a small tennis ball inside the car. As far as your friend is concerned, the ball dropped straight down, but because the car is moving, you see the ball fall in a curved path. You and your friend, in other words, give different descriptions of the event. There is actually nothing surprising about this, because the two of you are in different frames of reference.

Now, however, suppose I asked what the basic laws of physics were in your respective frames of reference, with the assumption that you both had sufficient laboratory facilities to find an answer to my question. As far as Newtonian mechanics is concerned, you would both find the same laws operating. The statement that the laws of mechanics are the same in all frames of reference is known as the principle of relativity. In suitably expanded form, it is the basis for general relativity, our best theory of gravitation and our current effort to understand black holes.

The second item isn't normally stated explicitly by people working in Newtonian mechanics, but it plays the role of a universal, though unstated, assumption. It has to do with the nature of time. In the example we just considered about your friend in the car, you'll notice that we never said anything about the behavior of the clocks you and your friend had. The reason for this is simple: we simply assumed that all clocks run at the same rate. A second is a second, after all. But if you think about it for a moment, you'll realize that when we do this we are making an assumption about the fundamental nature of time. We are assuming that there is a single correct time, a clock kept by God, or perhaps by the blind fish in the Caspian described in the epigraph. The assumption seems to match our experience, but questioning this unspoken assumption was one of the things that led Albert Einstein to the theory of relativity. In addition, for the Newtonians space was not a thing but simply the empty stage on which events took place.

Newton's work gave scientists a road map for exploring the universe. His laws of motion, as far as we could see, applied to every object in the universe, from galaxies to billiard balls. It wasn't long before scientists began to apply Newton's reasoning to events involving other things.

## THE ELECTROMAGNETIC WORLD

We continue our story by investigating an unexpected result that came out of applying Newton's methods to the investigation of the phenomena of electricity and magnetism. Both of these phenomena had been known since antiquity; indeed, the Chinese invented the first compass around 200 BCE, and the Greeks investigated static electricity in 600 BCE. It wasn't until the nineteenth century, however, that it was discovered that these two seemingly disparate phenomena were related. Essentially they were two sides of the same coin. The invention of the battery allowed scientists to discover that moving electrical charges, or what we call electrical currents, would create magnetic fields and that changing magnetic fields could produce electrical currents.

These discoveries, made by many different scientists, were brought together in 1873 by the Scottish physicist James Clerk Maxwell (1831–79), who added a term that had been missing up to that point. What he showed was that all the phenomena associated with electricity and magnetism, as well as the connections between them, were summarized in a series of four equations, now universally known as Maxwell's equations.

Maxwell was an expert on the advanced math of his day. Think of him as being analogous to someone who can navigate around Big Data today. Using techniques from a field that we call partial differential equations, he showed that the basic equations of electromagnetism predicted the existence of a strange kind of wave composed of electric and magnetic fields. Basically, the magnetic fields changed, which produced an electric field, which in turn produced a changing magnetic field, which produced another electrical field, and so on. Think of the energy in the wave being tossed back and forth between the electric and magnetic fields as the wave moves along.

What was really interesting was the speed of this strange wave depended on the forces between electrical charges and the forces between magnets. Both of the quantities were well known, and when Maxwell plugged in the numbers, he found that the waves would move at 186,000 miles per second.

That's the speed of light. What Maxwell's mathematical manipulations had proved, in other words, was that light was an aspect of electricity and magnetism. This is such an important result that this speed is customarily given its own letter of the alphabet, $c$. The realization that light is connected to the same processes that stick notes of your refrigerator and generate static cling was another of those great unifications that make up the history of science, like Newton's unification of earthly and heavenly gravity.

It's an interesting fact that, like Newton, Maxwell had a hard time imagining a wave that could travel through a vacuum. Like all scientists at the time, he assumed that space was filled with a tenuous substance, called "luminiferous ether," and he devised complex models (gears, sprockets, and all) to explain how his waves worked. Today we understand that light can move along just by exchanging electric and magnetic fields, without needing any sort of medium to support it.

One interesting aspect of electromagnetic radiation is that, according to the equations, it should come at all wavelengths. The wavelengths of visible light range from about four thousand atoms long (for violet light) to about eight thousand atoms long (for red light). For Maxwell, most of what we call today the electromagnetic spectrum was missing. One author compared the situation to seeing a full orchestra on a stage but hearing only the piccolo. Beginning in 1888, when Heinrich Hertz discovered radio waves, the entire panoply of wavelengths, from radio waves (with wavelengths measured in miles) to gamma rays (with wavelengths shorter than the nuclei of atoms), had been found. All of these waves move with velocity $c$ in the vacuum.

With a full understanding of Maxwell's equations, then, another corner of the universe was brought inside the purview of science. Just as Newton's laws gave us the tools to understand motion wherever it occurred, Maxwell's equations gave us the tools to understand electrical and magnetic phenomena. There was, however, an important difference between these two great unifications. There is a specific speed—the speed of light—built into Maxwell's equations, but there is no such speed built into Newton's laws. It was this fundamental difference that Albert Einstein exploited in the early years of the twentieth century.

Note that this fundamental difference has an extremely important consequence. If the laws of relativity are going to tell us that the laws of nature have to be the same in all reference frames, then the speed of light has to be a universal constant, or the same everywhere. Otherwise the laws of nature, and specifically Maxwell's equations, are going to depend on the observer's frame of reference.

## EINSTEIN AND RELATIVITY

It's so seldom that a single individual has an enormous effect on the development of science that it's worth pausing for a moment to talk about the life of Albert Einstein, especially since it was his theory of general relativity that led scientists to the concept of the black hole.

Einstein was born in 1879 in the town of Ulm in southern Germany. His family was what would be characterized as upper middle class today, his father a partner in a manufacturing firm that produced feather beds. They were Jewish, but because Albert was not given a biblical name, scholars believe they were not observant. He was apparently slow to start speaking, though in later life he said he didn't want to start talking until he could speak in complete sentences. In 1880 the feather bed business collapsed, and the family moved to Munich, where, at the age of six, Albert started school.

An urban legend has it that there he flunked grade school math, but in fact, we have letters from his mother to his grandmother bragging about what a good student he was. After finishing gymnasium (roughly the equivalent of high school) he applied for admission to the Eidgenössiche Technische Hochschule (ETH) in Zurich, then as now one of Europe's leading technical universities. It was here that the legend of Einstein as a poor student got its start, because he flunked the entrance exam. It wasn't the physics and math that did him in, though. It was the fact that the exam included subjects like drawing and literary history, which he hadn't mastered.

He was eventually admitted to the ETH, although his attitude rubbed some of his professors the wrong way. He was characterized as a young man who was reluctant to accept advice from his elders, a trait not uncommon among college students. When he graduated in 1900, he landed a couple of temporary teaching appointments that kept him going until the father of a college friend introduced him to the head of the Swiss Patent Office, and in 1902 he became a patent examiner third class.

Einstein would make his mark in the esoteric field of theoretical physics, but before that he was a great, eminently practical patent examiner. He seemed to have had a real feeling for mechanical apparatus, and during his lifetime had no fewer than fifty patents issued in his own name. Perhaps the most well-known of these is U.S. Patent 1,781,541, issued in 1930, protecting Einstein's solution to a serious problem that developed when people began installing refrigerators in their homes. The mechanical pumps in the refrigerators were often leaky, releasing toxic gases into the home's environment. Einstein's patent, submitted together with physicist Leo Szilard, replaced the mechanical pumps with a sealed compressor that had no moving parts but did its work with electromagnetic fields. The device was never commercialized, but it illustrates one legacy of his time in the Patent Office.

All through his time as a patent examiner, Einstein kept publishing papers in theoretical physics. In 1905, which physicists still refer to as the *annus mirabilis* or miracle year, he published four important papers, any one of which could have resulted in a Nobel Prize. From our point of view, the most important of these introduced the theory of special relativity. The other papers deal with the photoelectric effect, Brownian motion, and mass-energy equivalence. The first of these was one of the seminal papers in the development of quantum mechanics and the work for which he was awarded the Nobel Prize in 1923; the second provided convincing evidence for the reality of atoms; and the third led to the famous equation $E = mc^2$. For our purposes, however, it is special relativity that needs attention.

According to Einstein, the idea for relativity came to him one day when he was riding in a streetcar in Bern, Switzerland. Looking at a nearby tower that had a large clock on display, he realized that if the streetcar were to move away from the tower at the speed of light, it would appear to him that the clock had stopped. It seemed, in other words, that what an observer saw as the reading on the clock would depend on the observer's frame of reference. The faster the observer was moving, the more the clock would appear to slow down.

Incidentally, this method of trying to understand some natural phenomenon by thinking about an imaginary situation is customarily described as performing a *Gedanken* (from the German *denken*, "to think") or thought experiment. It was one of Einstein's favorite techniques. It isn't necessary to perform the experiment; Einstein didn't have to accelerate his streetcar to the speed of light, for example. He could learn what he had to learn just by thinking about doing so.

The point is that if different observers in different frames of reference see the same clock ticking at different rates, we have to call into question the unquestioned Newtonian assumption about the existence of a universal time. Time, like space, becomes a variable that depends on the observer's frame of reference, a fact that explains why you will often see the word "space-time" in the rest of this book.

There is, however, a deeper issue involved in the world as seen from different frames of reference, an issue that arises from the fact that the speed of light is built into Maxwell's equations. To see this, go back to the example of your friend dropping a ball in a moving car while you watch from the side of the road. Suppose now that your friend, instead of dropping the ball, throws it forward. Suppose further that you, standing next to the road, see a car going by with a velocity $v$, and suppose your friend inside the car sees the ball moving forward with velocity $u$. Now perform the Gedanken experiment of measuring the speed of the ball that you, on the ground, see. Our intuitive answer is that you will see the ball moving with velocity $v + u$, the speed of the ball plus the speed of the car.

Now suppose we perform a slightly different Gedanken experiment. Suppose that instead of throwing a ball your friend shines a flashlight, and the two of you measure the speed of the emitted light. It's pretty obvious that your friend will see the light moving with the customary speed $c$, but what about you? Will you see the light moving at a velocity $c + v$?

This is where the world of electromagnetism enters the picture, because, as we stressed earlier, the speed of light is built into Maxwell's equations. If your friend sees the light moving with velocity $c$ and you see it moving with $c + v$, then Maxwell's equations will be different in your two frames of reference. This would mean that the laws of electromagnetism would be different in the two frames, and this in turn would mean that the principle of relativity would not apply in the world of electromagnetism as it does in the world of mechanical motion. The only way to find out if the principle is universal is by experimentation. There are several ways that Maxwell's equations could be modified to allow light to have different speeds in different frames of reference while preserving something like the principle of relativity:

1. The $c$ in the equations might refer only to the speed of light in the ether.
2. There might be terms in the Maxwell equations, as yet undetected, that make the speed of light dependent on the speed of the source.

3. There may be something wrong with the way we think about space and time when we calculate velocities.

The first of these options was already pretty much a dead letter by the time Einstein began his work on relativity. In 1887 physicists Albert Michelson and Edward Morley, working at what is now Case Western Reserve University in Cleveland, had already shown that the ether, so beloved of ancient philosophers, simply didn't exist. They reasoned that if there was an ether and Earth was moving through it, there should be a detectable ether wind, much as there is a perceptible wind in a moving vehicle on a still day. The only conclusion we can draw from the fact that Michelson and Morley (and generations of experimentalists after them) failed to detect the ether wind is that there is no ether to produce it. And although Einstein later said that this experiment did not influence his thinking, it certainly allows us to reject option 1.

Option 2 can also be eliminated using some basic astronomy. It happens that many of the stars we see in the sky are actually double star systems—systems in which the stars rotate around each other. Light from a star in such a system will be emitted when the star is moving toward you part of the time and away from you the rest of the time. It's a simple matter to detect a change like this, and from the fact that we don't see any such effect we have to conclude that there are no extra undetected terms in Maxwell's equations.

That leaves us with option 3—the one that Einstein chose to investigate. After all, our arguments involve comparing velocities, and velocity is defined as distance divided by time. If our understanding of distance (that is, space) and time are wrong, nothing that we say based on our incorrect concepts can be trusted.

Before we plunge into the theory of relativity, however, there is one point we have to make. Nothing that we have done up to this point *proves* that the principle of relativity is correct. What we will do is look at the predictions that relativity makes and see if they match what nature shows us. This is the only way to justify adopting any new theory.

Einstein's ride on that streetcar gives us a hint about how to proceed. The fact that the time shown by the clocktower depends on the velocity of the observer tells us we should begin our investigations by looking at time and motion. Furthermore, if we are going to take the principle of relativity seriously and assume that it applies to all laws of nature, we are going to have to assume that the speed of light, $c$, is the same in all frames of reference. Without this assumption, Maxwell's equations would be different in different frames of reference.

This is, as we shall see, a huge assumption—so huge, in fact, that it often listed as a separate assumption in discussions of relativity. However you want to think about it, it is an essential part of what follows.

One thing we have to realize about time is that physicists do not consider it our job to say *what* it is, but only to find a way of measuring it. Traditionally, we do this by finding a regularly repeating event, use it to define a basic unit of time, and then measure all other times in terms of that basic unit. Nature, in fact, gives us several such events. We list some of them below along with the regularly repeating event associated with it.

- The year—the time it takes Earth to complete an orbit around the sun
- The day—the time it takes Earth to complete one rotation on its axis
- The month (approximately)—the time it takes the Moon to complete an orbit around Earth

Starting in the Middle Ages, humans began adding items to the list of regularly repeating events to define the basic unit of time. The swinging of a pendulum was an early choice. Since 1967, the second has been defined in terms of the energy levels of electrons in a cesium atom, a definition that we still use today.

All of this tells us there is nothing preventing us from imagining a Gedanken experiment that uses a new kind of regularly repeating event. For this experiment we will construct in our minds a device known as a light clock. It has three parts: a flashbulb that generates a beam of light, a mirror that reflects the beam back toward the flashbulb, and a photocell that registers the arrival of the beam and triggers the flashbulb, thereby starting the whole process again. The "ticking" of this clock would then be "flash, reflect, click at the photocell."

Remember that according to the rules of Gedanken experiments, we don't have to actually build this clock, we just have to think about it. Another point that may help you visualize the operation of the light clock is to note that it can be synchronized with any familiar clock you want. Do you like to picture a grandfather's clock with a swinging pendulum? Fine—just move the mirror far enough away from the flashbulb so that it takes a full swing of the pendulum for the light to get back. This would involve placing the mirror most of the way to the Moon. (Are you starting to see why not having to build the apparatus in a Gedanken experiment is such an advantage?) The point, however, is that anything we prove about a light click will be true of any clock you can imagine.

Now for the experiment. Let's start with two identical light clocks, one on the ground and the other on an open railroad car moving at a constant velocity with respect to the ground. Let's further arrange things so that the two flashbulbs both go off as the moving clock passes the clock on the ground. What will the observer on the ground see?

The ground observer will see his or her clock behaving normally—the light travels straight up to the mirror and comes straight back down again. Looking at the moving clock, however, he or she sees something else entirely. While the light in the moving clock is traveling up to the mirror, the train will move. Thus, the ground observer sees the light in the moving path traveling on a slanting path. The same thing happens as the light in the moving clock travels down to the photocell. The end result is that the ground observer seeing the light in the moving clock traveling in a sawtooth pattern.

This in turn means that as far as the observer on the ground is concerned, the light in the moving clock has to travel farther between ticks than the clock on the ground. And this is where the principle of relativity comes in, because if the speed of light is the same in both frames of reference, the photocell on the ground will go off before the photocell on the moving train. In other worlds, according to the principle of relativity, *moving clocks slow down!*

Not just light clocks. Every clock. This conclusion, known as "time dilation," is one of the fundamental results predicted by the principle of relativity. Without going into a lot of detail, we can say that this prediction has been repeatedly confirmed by experimentation since Einstein published it in 1905. Perhaps the most dramatic confirmation occurred in 1971, when atomic clocks were flown on commercial airliners traveling around the world and compared to clocks that had remained stationary. The results were exactly as Einstein had predicted.

Further verification of time dilation comes from experiments on cosmic rays. They produce sprays of unstable particles high in the atmosphere, particles that should decay before reaching the ground if it weren't for the effects of time dilation. The fact that they do make it to the ground means an observer sees that the moving particles have slowed down. Other arguments, only slightly more complex than those given earlier, result in other (well-verified) predictions: moving objects get more massive and shrink in the direction of motion, and then there's the famous mass-energy equivalence $E = mc^2$.

All of these predictions come out of arguments, like the one for time, based on clocks and meter sticks moving at a constant velocity— no acceleration or rotation. These are called "inertial" frames of reference, and the field that deals with them is known as special relativity.

Special relativity is simple, requiring math no more complex than basic algebra and the Pythagorean theorem. It is routinely taught to thousands of physics undergraduates every year.

The basic statement of special relativity is that the laws of nature, including Maxwell's equations, are the same in all inertial frames of reference. To go from this to the basic statement of general relativity—the statement that the laws of nature are the same in *all* frames of reference—is a difficult task. For one thing, the math becomes much more complicated. (It took Einstein ten years to master it.) One source of the difficulty is that when we deal with space-time we are looking at a four-dimensional system. There are three space dimensions (up-down, front-back, left-right) and one for time. Thus, in their full glory, Einstein's results come out to sixteen coupled partial differential equations. which is customarily reduced to only ten coupled partial differential equations because of inherent symmetries in the system. No wonder it took him so long to get there.

Complex or not, though, we're going to have to have some sense of what general relativity is about before we tackle black holes and the origin of the universe. Fortunately, our old friend the Gedanken experiment will help.

# 2

# IS GRAVITY REALLY
# A FORCE?

———

## Things are seldom what they seem.

*W. S. Gilbert and Arthur Sullivan,* H.M.S. Pinafore;
or, The Lass That Loved a Sailor

**W**e have seen that the principle of relativity can be used to gain insights into the nature of time in a world that contains only inertial frames of reference. Similar results give us what we call the special theory of relativity. We also pointed out that extending the principle of relativity to all frames of reference—leading to what is called the general theory of relativity—involves some pretty complex mathematics. In what follows we will look at some simple examples of how the general theory works and, in the process, raise a fundamental question about the nature of gravity.

Let's start with a familiar noninertial frame of reference, the rotating Earth. For our first Gedanken experiment we will have someone standing on the North Pole, throwing a ball to someone standing on the equator. We will look at this process from two frames of reference—one in a stationary spaceship high above Earth, the other located next to the catcher at the equator. What do these observers see?

The observer in the spaceship sees the ball moving in a straight line, as required by Newton's laws of motion, with Earth rotating under it as it flies southward. During the flight, the only force acting on the ball will be Earth's gravity pulling the ball downward and keeping it from flying off into space. This observer will see no forces acting in the horizontal direction.

The observer on Earth sees something entirely different. As soon as the ball leaves the pole, it begins to deviate from a straight line, eventually landing on a spot behind the catcher—actually the spot the catcher occupied when the ball was thrown. According to Newton's laws of motion, the fact that the ball does not move in a straight line means that there must be a force acting on it. In fact, we call this the Coriolis force, after the French engineer and mathematician Gustave Coriolis (1792–1843), who first analyzed the effects of rotation on motion.

This is a strange situation. We have a "force" that is present in one frame of reference but not in another. Physicists long ago learned to deal with this kind of situation and refer to things like the Coriolis force as "fictitious forces." Clearly the fictitious forces are connected to the accelerated frames of reference in which they are seen. They are also somehow connected to the fact that there are two obvious ways to observe the system: from inside and outside.

We can emphasize this fact by pointing out that there is a third way to analyze our Gedanken experiment, one that we don't normally think about but that is very much in keeping with the way we'll get to general relativity. In our two ways of looking at the experiment, we have (implicitly) assumed that the coordinate system on Earth is not affected by the planet's rotation; it simply spins along with the solid ground. But what if we changed that assumption? Suppose we said that the rotation warped the coordinate grid on the planet so that the equivalent of lines of longitude became curved in such a way that, to an observer on Earth, they marked out the path that a terrestrial observer would see the ball follow. (The technical term for such a path is "geodesic.")

In this case an observer on Earth would say that the presence of rotation warped the coordinate grid, and that in the new grid that ball follows a geodesic line. In this case we wouldn't have to introduce fictitious forces; we'd just have to modify Newton's laws to say that objects move along geodesics in the space that has been warped by the rotation.

Once we have the idea that we can analyze motion by changing the coordinate system, we can move on to a second set of Gedanken experiments. Suppose we have a ship in deep space whose occupants can't look outside. Let's start by having the ship moving with a constant velocity and have an astronaut hold a ball out at arm's length and let it go. An observer outside the ship will see the ball continue to move at the same velocity as the astronaut, and the astronaut will see the ball float. No fictitious forces arise in either frame of reference.

Now suppose we have the spaceship accelerating upward. When the astronaut lets go of the ball it will continue to move at whatever velocity it had when it was released. An outside observer will see the floor of the spaceship accelerate and move upward until it encounters the ball. In this case we see no forces acting on the ball. The astronaut in the ship, on the other hand, will see the ball fall toward the floor, eventually bouncing off of it. He or she would have to conclude that there was a force acting to pull the ball down. Like Newton and his apple, the astronaut would call that force gravity.

But, of course, this is just how we defined a fictitious force, one that is present in one frame of reference but not in another. To make this point more strongly, imagine that the spaceship is accelerating upward at a rate of 32 feet per second per second, the acceleration due to gravity at Earth's surface. There is no way the astronaut could tell if the ship were accelerating in deep space or sitting on the ground at Cape Canaveral. Gravity, in other words, may well be a fictitious force, just like the Coriolis force!

Just as we discussed a "third way" to look at the Coriolis force, a way that involved allowing the rotation of Earth to distort the coordinate grid on the planet, we can think of allowing a massive body like Earth to warp the structure of space-time near its surface so that the geodesic paths are those corresponding to a free fall under the influence of gravity. We can furthermore generalize this process to any massive body by applying a two-step argument:

• The presence of mass distorts the fabric of space-time.
• Objects move along geodesic lines in that distorted space-time.

This in fact is a simple summary of what Einstein found when he applied the principle of relativity to all frames of reference. It is the heart of general relativity and our current best theory of gravitation.

A common way of picturing general relativity is to imagine a rubber sheet stretched taut with a set of lines marking out a square coordinate system. Now imagine dropping a bowling ball in the middle of the sheet. The sheet will be distorted, as space-time is distorted in our earlier examples. Now roll a marble toward the bowling ball. The marble will be deflected as it comes near the bowling ball. A Newtonian physicist would say that the deflection is caused by the force of gravity pulling on the marble. A general relativist would say that the space-time coordinate system had been distorted by the presence of the earth and that the marble was simply following a geodesic path in that distorted coordinate system.

Same result, different interpretations.

Let's pursue our Gedanken experiments one more time. Suppose that in the accelerating spaceship the astronaut throws the ball toward the wall. Since the ship will accelerate between the time the ball leaves the astronaut's hand and the time it hits the wall, the astronaut will see the ball moving downward, hitting the wall below the place it would have hit in the absence of acceleration. The astronaut would attribute that drop to the action of gravity.

Suppose, however, that the astronaut shines a flashlight toward the wall. Where will the beam of light hit? Given that the speed of light is an invariant, the light will hit the wall below the spot where we would expect it to hit in the absence of acceleration. In other words, general relativity predicts that massive objects will deflect light rays that pass near them. It was the detection of this sort of deflection that the great mathematician Sir Arthur Eddington (1882–1944) measured during a solar eclipse in 1919 of light passing near the Sun. His results provided experimental confirmation of general relativity and vaulted Einstein into international fame. (Newtonian gravity also predicts that light rays will be bent near a massive object, but it predicts only half the deflection associated with relativity and measured by Eddington.)

Physicists were inclined to accept general relativity because the theory had already resolved an old problem with the orbit of Mercury. Planetary orbits are elliptical in shape, and because of the gravitational effects generated by the other planets, the point in Mercury's orbit when it was closest to the Sun moved around. This phenomenon is known as precession. Most of the precession of Mercury's orbit could be explained by applying ordinary Newtonian gravity, but there had always been a tiny discrepancy between the Newtonian predictions and the data. It turned out that general relativity supplied the small missing piece needed to resolve this problem.

# 3

# THE BIRTH OF
# BLACK HOLES

### Then it's hi! hi! hee!
### In the field artillery,
### Shout out your numbers loud and strong

"The Caissons Go Rolling Along"

I t was cold on the Russian front that day in 1915. The German artillery officer stamped his feet and wrapped his greatcoat around him as he looked at the strange package that had arrived in the battalion mailbox, wrapped in butcher's paper and tied with string. When he opened the package later in his billet, he was surprised to see that it was an issue of the Prussian Academy of Sciences' journal containing a paper written by his erstwhile colleague in Berlin, Albert Einstein. This was the famous paper in which Einstein laid out general relativity.

The man was no ordinary artillery officer. Before the start of the First World War Lieutenant Karl Schwarzschild had been Herr Doktor

Professor Karl Schwarzschild, the director of the prestigious Potsdam Astrophysical Observatory, one of the most important positions in German science. Because of his age and position he could easily have avoided military service, but like many European men caught up in the nationalistic fever sweeping the continent, he volunteered for the army. He spent the early years of the war working out the effects of wind and temperature on the trajectory of artillery shells, a task that brought him to the Russian front on that fateful day in 1915.

People who had wrestled with Einstein's equations up to this time had tried to in general solve them for the entire universe. Schwarzschild took a different path, one well known to theoretical physicists. Confronted with a complex system that defies analysis, one way to proceed is to look for the simplest possible example of that system in operation. The hope, of course, is that the solution for the simple case will give you a hint about dealing with the more complex one.

Following this path, Schwarzschild decided to look at a system consisting of a single massive body surrounded by empty space. Think of this as an attempt to apply general relativity to a single isolated star. The advantages of this approach are obvious. For one thing, you don't have to worry about the direction in which you approached the massive body— you can just assume it's a sphere so that all directions are the same. In addition, by taking this approach you can ignore things like the effects of electric charge and rotation, complications that would have to be dealt with eventually but could be ignored at the start.

With these simplifying conditions, Schwarzschild was able to solve the Einstein equations, laying out simple results that described the warping of space-time by the presence of an isolated mass. His solution had many appealing properties. At large distances from the mass, for example, it gave the same well verified results that Isaac Newton had written down centuries before. It was only as you got close to the mass that things began to get weird.

Schwarzschild approached this weirdness by thinking about the concept of escape velocity. Look at it this way: when you throw a ball up from the surface of the Earth, it can climb to only a certain height before gravity makes it slow down, stop, and fall back down. There is some initial velocity, however, that will be so fast that the earth won't be able to hold on to the ball. It will just keep traveling upward and leave the planet forever. We call this transitional speed the "escape velocity." For Earth, it's about seven miles per second.

As a seasoned artillery officer, Schwarzschild would naturally think of objects traveling near his central mass in terms of escape velocity. Put an object anywhere in space around the central mass, and there will be some outward velocity that will allow it to escape and find refuge in the Newtonian outer reaches of the system. He saw that the closer you got to the central mass, the higher the escape velocity would be.

It was at this point Schwarzschild noticed something strange. If you look at the escape velocity as objects get closer and closer to this central mass, you eventually reach a point at which the escape velocity exceeded the speed of light. Since one of the main results of the theory of relativity is that nothing can travel faster than light, it followed that nothing closer to the central mass than this crucial distance can ever escape to the outside world. That fact defines what we now call the "event horizon"—the distance of closest approach beyond which escape becomes impossible. Though he didn't know it at the time, Schwarzschild had discovered what would later be called a "black hole."

We don't have direct evidence that Schwarzschild and Einstein had known each other in the prewar days, but as prominent members of the Berlin scientific community they would certainly been aware of each other. Schwarzschild sent his results to Einstein with a friendly letter, ending with, "As you see, the war is kindly disposed toward me, allowing me, despite gunfire at decidedly terrestrial distance, to take this walk into your land of ideas."

Einstein was impressed by Schwarzschild's work; he said he had had no idea that such a simple solution to his equations was possible. Consequently, he presented the work to the Prussian Academy, in those days standard procedure for publicizing work that someone else had done.

We would love to have been able to close this excursion into history by recounting Schwarzschild's triumphant return to a position of leadership in the German scientific community. Alas, it was not to be. A few months after Einstein's announcement, Schwarzschild was diagnosed with a disease known as pemphigus, an autoimmune condition often found among Ashkenazi Jews. He was invalided out of the army and died shortly thereafter in Germany,

Perhaps the most amazing thing about Schwarzschild's results as far as black holes are concerned is that for almost half a century no one believed that his results could apply to objects in the real world.

## CLASSIC BLACK HOLES

The key fact about Schwarzschild's solution to the Einstein equations is that when an object gets closer than a certain distance from the central mass it can never get back out. This distance is called the Schwarzschild radius, and the spherical surface it defines in space is called the event horizon. The Schwarzschild radius of an object depends on its mass. The higher the mass, the greater the Schwarzschild radius. Put another way, for every object of a given mass, there is a radius defined by the fact that if all of the mass of the object is confined within that radius, that object will become a black hole. The density of mass in a black hole can be extraordinary. To turn Earth into a black hole, for example, the entire mass of the planet would have to be confined within something about the size of a marble.

Once we have established the existence of an event horizon, however, we can start to think about what it would be like to approach a black hole. Suppose you had three spaceships located in deep space, far from the black hole you want to approach. Each of these spaceships is equipped with a device that sends out light signals of a given frequency and another device capable of detecting light of all wavelengths. Two of the spaceships set out side by side toward the black hole, while the third stays in deep space and watches. What will observers on the spaceships see?

To answer this question, we have to go back to one more prediction of general relativity, one involving the interaction between light and gravity. Imagine shining a flashlight straight up from the surface of the earth. One way of thinking about what happens is to say that the light is climbing up from the bottom of a gravitational hole. It takes energy to do this. If we threw a baseball, we know that it would slow down as it climbed up. But light *can't* slow down—it must travel at the velocity we've called $c$, so the energy has to come from somewhere else. According to general relativity, as the light climbs up its frequency (and hence its energy), blue light shifts down to red, which shifts down to infrared, and so on. This is called the "gravitational red shift." It is one of what are called the classic tests of general relativity, which was first seen in the laboratory in 1959 and today is universally used in GPS systems, which involve clocks high above Earth sending signals up and down in Earth's gravitational field.

Going back to our spaceships, we can think of the light being emitted from each as a clock that is ticking a billion billion times per second. As the twin spaceships descend toward the black hole, they will move faster and faster, and as a result the time dilation will get bigger. This will be interpreted by the distant observer as the clocks on the traveling ships slowing down. As far as this observer is concerned, the clocks on the falling ships will "tick" more and more slowly until, when those ships reach the event horizon, they stop entirely.

On the other hand, the observers in the moving spaceships see their own clocks and the clocks in their companion ship ticking normally. For them there is nothing special about the event horizon—they just sail right on by.

We can carry the spaceship analogy a little farther by asking what happens once they are inside the event horizon. The passengers on the spaceships won't notice anything in particular when they pass inside their "black hole": the same laws of physics will apply both inside and outside. There will come a point, though, when the spaceships will reach what is called a singularity at the center of the black hole. It is a point when the familiar rules of physics break down and all hell breaks loose. The existence of the singularity was one of the main reasons physicists refused to accept the reality of black holes in the early and mid-twentieth century.

Black holes showed up in the physics literature early on, though they weren't called black holes at the time. For a generation of scientists who were used to exploring the heavens with telescopes, black holes posed special problems. We normally detect objects in the heavens either by the light they emit (we see stars this way) or by light they reflect (we do this every time we look at the moon). Neither of these will work for a black hole, because once light passes inside the event horizon it can never come out to enter your telescope.

But that wasn't the only problem. As we have pointed out, according to Schwarzschild's solution, every object in the universe has a Schwarzschild radius and could therefore become a black hole, at least in principle. All that has to happen is that all of the mass of that body has to be crammed inside that radius. The problem was that no one could imagine any physical process that could produce that kind of compression. As we have seen, for Earth to fit inside its own event horizon the planet would have to be crushed down until it was about the size of a marble.

The small size of black holes relative to their mass gives rise to an interesting phenomenon that wasn't noticed by Schwarzschild and his colleagues but has become part of the folklore of black holes. You can understand this phenomenon by recalling that the force of gravity between two bodies increases as the bodies get closer to each other. Cut the distance between the bodies in half and the gravitational force will increase by a factor of four. Cut it to a third and the force increases by a factor of nine, and so on. The closer you can get to the center of a gravitating body, therefore, the greater will be the gravitational force you experience.

When you are on Earth's surface you are about four thousand miles from the center of the planet, and the force of gravity at that distance is what produces your weight. Were Earth to become a black hole, however, you would be able to stand a fraction of an inch from the center. At this distance, your weight would increase over a million trillion times over what it is now.

This raises an interesting question about what would happen to a human observer who tried to approach a black hole. Obviously the force of gravity on the observer would increase as he or she approached the event horizon. Long before arrival, however, another effect would make itself felt. You can understand this new effect by noting that when you stand on Earth's surface your feet are closer to the planetary center than your head. This means that the gravitational force on your feet is (infinitesimally) larger than the force on your head. The extra few feet your height adds to Earth's radius is so small that it can safely be ignored. If Earth became a black hole, however, the force associated with those extra few feet would become huge.

Suppose our hypothetical traveler was approaching the compressed Earth's event horizon feet first. The difference in gravitational attraction between his or her head and feet can be thought of as a force acting to stretch the observer's body. As the observer approaches the event horizon, that stretching force would become larger and larger until, eventually, it exceeded the internal forces holding the body together. Most likely the ligaments holding the joints together would snap first, but eventually the difference in gravitational force would become so great that the bones themselves would be pulled apart. This stretching phenomenon has been given the amusing name of "spaghettification." It can be a hazard for astronauts who get too close to a black hole. Having said this, however, we will argue later that objects approaching a supermassive black hole will not be spaghettified. This is because the supermassive's radius is large enough to cancel the effect of its large mass.

Assuming the observer has survived the journey though the event horizon, we can use Schwarzschild's solution to talk about the rest of the journey, even though we can no longer communicate with the inside of a black hole. At the very center of the interior of a black hole, Schwarzschild's solutions simply blow up, a term physicists use to describe the process by which quantities dealt with in their equations become infinite. Since nothing in nature can actually become infinite, physicists preferred to believe that black holes couldn't actually exist rather than face the fact that their theories didn't work. It was an easier way out of the problem than the proposed solutions we'll talk later.

## A BLACK HOLE MENAGERIE

Once theorists accepted the reality of black holes, they were free to apply their theories to the question of the kinds of black holes that might be out there. Not surprisingly, over the years, many different types of black holes have been proposed, some of which have actually been discovered, some of which have not (and may never be). Here's a sampling of the modern black hole bestiary.

## Quantum Black Holes

Quantum black holes are hypothetical objects whose existence has been suggested but which have never been seen in nature. They are supposed to be black holes the size of elementary particles. They may have been created in the early stages of the Big Bang, when there was a lot of loose energy floating around. They may also have been created in collisions of high energy cosmic rays, although this is unlikely.

In any case, there is a process known as Hawking radiation that causes black holes to evaporate over time. The smaller the black hole, the faster the evaporation process. Even if quantum black holes were created in the Big Bang, then the chances are that they would have evaporated by now.

One interesting bit of silliness concerning quantum black holes occurred in 2008, when the Large Hadron Collider—the world's largest particle accelerator—was going to be turned on in Geneva, Switzerland. In this machine, protons are accelerated to close to the speed of light and then allowed to collide head on. A group of activists argued that these collisions would create black holes that would consume the earth, and they went to federal court (in Hawaii, of all places) to prevent scientists from turning the machine on. The judge, thankfully, listened to scientists who pointed out that cosmic rays of much higher energy than those available at CERN had been bombarding Earth for 4.5 billion years without creating any black holes.

## Primordial Black Holes

Primordial black holes are another type of black hole that have been talked about but never seen. Theorists have proposed the category because recent discoveries by the James Webb Space Telescope have indicated that supermassive black holes, about which more later, are present in the earliest galaxies we can see. This naturally raises the question of how those black holes got so big so fast. If some large black holes formed by some as yet unknown process in the Big Bang, it could answer this question.

## Stellar Black Holes

Stellar black holes, having a mass typical of large stars, were the first to have been discovered. To understand how these black holes came to be we need to take a small astrophysical excursion and discuss how stars work.

Stars begin their life in the gas clouds that litter space. These clouds are lumpy in the sense that some regions have more mass than others. The extra gravitational attraction associated with that mass pulls more matter in, which increases the mass, which increases the gravitational attraction, and so on. Eventually that original lump becomes a massive sphere of gas being pulled together by its own gravity. The sphere heats up, and when it gets hot enough it initiates nuclear fusion reactions that turn hydrogen into helium, generating energy in the process. This energy streams outward, creating a pressure that balances the force of gravity. The star stabilizes and for the rest of its life burns its nuclear fuel to fight off the inward pull of gravity. The light that you see when you look at a star is what leaks out after the star has used the energy to fight off gravity. Think of it as the by-product of an epic battle.

Eventually the star loses. It runs out of fuel, while gravity just never quits. Depending on the mass of the star, there are a number of quantum mechanical processes that can prevent total collapse. For our purposes, we simply note that if the mass of the star is more than ten times the mass of the Sun, all of the material of the star will be pulled inside its Schwarzschild radius, and a stellar black hole will be born.

The historical debate about the existence of black holes was centered on this process, and consequently on the existence of stellar black holes. We should also point that it was the discovery of a stellar black hole that eventually proved that black holes actually exist.

## Intermediate Mass Black Holes

Suppose you were doing a study of the age of people in a given population, and you discovered that there were a lot of people under twenty
and a lot of people over fifty, but no one in between. You would immediately suspect that you were missing a part of the population and
begin searching for thirty- and forty-year-olds.

Our situation concerning the masses of black holes is similar. We
know of many stellar black holes with masses a few dozen times the
mass of the Sun and, as we shall see in the rest of this book, we know of
many black holes with masses of millions and even billions of times
the mass of the Sun. The search for masses in between is just starting.

A research team headed by Shobita, using the James Webb Space
Telescope, has found what may be an intermediate black hole in a
nearby dwarf galaxy.

## Supermassive Black Holes

No one expected this, but it turns out that at the heart of nearly every
normal galaxy there is a black hole with a mass millions of times that of
the Sun. We do not know how these black holes form, nor do we know
the role they play in the formation and structure of galaxies. They are the
most mysterious members of the black hole entourage, which is in and
of itself the most mysterious collection of objects in the universe.

This raises an important question. We know that galaxies are the
basic building blocks of the universe. We also know, thanks to the
James Webb Space Telescope, that galaxies formed early in the life of
the universe. At the moment we do not understand how large galaxies
and their attendant supermassive black holes formed as quickly as
they did. Thus, the study of supermassive black holes is intimately tied
to understanding the processes by which the universe formed. The
problem, of course—one we'll be returning to repeatedly in this book—
is that we still have a lot to learn about both the formation of galaxies

and the creation of supermassive black holes. Nevertheless, under-standing the formation of the universe remains one of the deepest questions that modern scientists investigate.

We have known since that day in 1915 that black holes are pre-dicted by general relativity. Since the discovery of supermassive black holes in the late twentieth century, we have learned that they are inti-mately tied to the formation and evolution of galaxies in the universe. It is unfortunate that Einstein died before any black holes were discovered and would never know the excitement his equations would generate.

# 4

# DO BLACK HOLES
# REALLY EXIST?

———

**I think there should be a law of Nature
to prevent a star from behaving
in this absurd way!**

*Sir Arthur Eddington on black holes,*
Royal Astronomical Society, *1935*

A s a physics graduate student at Stanford University in the late 1960s, Jim took a standard general relativity course. After working through Karl Schwarzschild's elegant solution of the Einstein equations, the professor pointed out the escape velocity issue and then launched into a long discussion designed to prove that while black holes were possible in theory, they could never actually form in the real world. This wasn't just some kooky minority opinion. In 1939 no less a scientist than Albert Einstein himself wrote a paper in which he said that the argument for the existence of black holes was "not convincing," and that they "could not exist in the real world."

Neither Einstein nor the Stanford professor actually used the term "black hole," of course. It was introduced in 1968 by American physicist John Wheeler and didn't come into popular usage for some time. Their arguments, though, reflected several aspects of the way people thought about (and rejected the reality of) black holes in the first half of the twentieth century. For one thing, scientists normally thought of black holes as being produced only by the collapse of stars—what we called stellar black holes in chapter 3. Furthermore, the arguments generally hinged on statements about the final state to which stars could evolve—a poorly researched subject in those days—and often depended on theories which are no longer accepted. The best way to understand the hiatus in black hole research from the 1930s to the 1960s and the reluctance of serious scientists to accept their existence is to briefly review what we know about the life cycle of stars.

## THE BIRTH AND DEATH OF STARS

Like human beings and all other life forms on Earth, stars are born, live out their lives following well-known rules, and eventually die. Stars are born in the great interstellar clouds scattered around the galaxy. Early in the lifetime of the universe these clouds were mainly composed of hydrogen gas created in the Big Bang, but as time went on, complex molecules, dust grains, and various heavy chemical elements were formed.

It is extremely unlikely that an interstellar cloud would be absolutely smooth throughout. In any cloud there will be regions where matter is more concentrated than it is in other regions—basically, the clouds will be lumpy. These heavy regions will exert a slightly stronger gravitational force on their surroundings than lighter regions, and hence will pull in matter and grow. This process makes them more massive and thus increases the gravitational force they exert on their surroundings. We expect, then, that the operation of the gravitational force will eventually cause the cloud to break up into massive clumps.

Gravity never quits. As a clump of matter separates out of the interstellar cloud, gravity will always be acting to pull the clump together. We can think of the life of a star as a series of attempts to neutralize the inward force of gravity—some attempts successful, others not.

As the material that will eventually become a star is pulled into a denser and smaller sphere, it heats up. The increased temperature results in the collisions between atoms and molecules in the cloud to become more violent, and eventually those collisions start tearing the electrons off of their nuclei. The result is that the protostar becomes a ball of what is known as a plasma, where negatively charged electrons and positively charged atomic nuclei whiz around, colliding with each other but not recombining.

Through all this process, gravity is pulling the protostar into a smaller and smaller sphere, a sphere that is continually heating up. Eventually, when the temperature at the center gets above a few million degrees Celsius, a new process starts. At this temperature the positively charged hydrogen nuclei are moving fast enough that their mutual electrical repulsion can be overcome, and the nuclei can come in contact with each other. Nuclear processes begin whose end result is to convert four hydrogen nuclei into the nucleus of a helium atom (two protons and two neutrons), along with a miscellaneous spray of particles and a lot of high-energy radiation. This process is known as nuclear fusion. The radiation from the core streams out, creating an outward pressure that counters the inward pull of gravity. The two forces balance, and the contraction stops. The outward streaming radiation eventually reaches the surface of the sphere and is released into space. The star begins to shine, and as long as the star has hydrogen to burn, the relentless inward pull of gravity can be neutralized. A star in this hydrogen burning state is referred to as a main sequence star.

The Sun began burning its hydrogen about 4.5 billion years ago and will keep burning it for another 5 billion years or so. Larger stars, because they have to work harder to overcome gravity, typically live for a much shorter time, perhaps some tens of millions of years, while smaller stars may burn their hydrogen so slowly that they live for trillions of years.

Eventually every star will run out of hydrogen fuel to stoke its nuclear fires. When this happens, the outward pressure that this fusion created drops, and the inevitable gravitational contraction resumes. The temperature in the core goes up and eventually reaches a point where the helium nuclei—the ashes of hydrogen burning—can come together in a fusion reaction that creates carbon nuclei (six protons and six neutrons). This process, in which the ashes of one nuclear fusion fire become the fuel for another, can go on for quite a while in massive stars. For a modest-sized star like the Sun, however, using fusion to create carbon is about it. Skipping over the details of a fairly complex process, stars like the Sun have to find a way to counter gravitational collapse that doesn't depend on nuclear fusion.

Enter the electrons. These were torn loose from their atoms early in the process of stellar formation and have been hanging around for billions of years, waiting to take center stage. Once the nuclear fires die down, the star will resume its gravitational contraction, squeezing the electrons closer and closer together. At this point we run into a stricture from quantum mechanics, the branch of science devoted to the behavior of the subatomic world. It turns out that electrons have an interesting property: two electrons can't be put into the same state in any system. You can think of this characteristic, called the Pauli exclusion principle, as a statement that electrons need elbow room—they can't be pushed together past a certain point. This means that the gravitational contraction will go on until the electrons can't be jammed together more. For a star like the Sun, this will happen when it is about the size of Earth.

The end state of the Sun will be a small, very hot star that is no longer generating energy. Stars like this are called white dwarves, and we can see them all over the place. Eventually, we expect, these stars will cool off and become cinders in space, or black dwarves.

More massive stars can burn through carbon to oxygen and all the way up the periodic table to iron (with fifty-six protons in its nucleus). Iron is the ultimate nuclear ash—you can't get energy from iron by any nuclear process, so massive stars have iron accumulating in their core. (Speaking of iron, the iron in your blood, the calcium in your bones, and the carbon in your tissues were all created in long-dead stars. As Joni Mitchell sang, we are all indeed stardust.) Eventually the loose electrons are forced into the protons in the iron nuclei, turning them into neutrons. Without the electrons to keep gravity at bay, the star's gravitational collapse resumes. The star explodes, throwing the chemical elements it has manufactured into space, where they join the interstellar clouds that will produce the next generation of stars.

This explosion, known as a supernova, allows the gravitational collapse to proceed until the core of the star, now made exclusively of neutrons, comes down to a size a few miles across. Because the neutrons, like electrons, need elbow room, they may stop the collapse and produce what is known as a neutron star in the midst of the gases and heavy atoms that were created in the supernova event. The upper limit of the size that can be supported by this sort of process is known as the Chandrasekhar limit.

If, however, the mass of the star is big enough—perhaps some tens of times the mass of the Sun—even the neutrons won't be able to stop the collapse, and the star will become a black hole. Our modern understanding is that there are several possible end states to stellar evolution: stars of different mass wind up in different states. This was not the way people saw things in the 1930s, a time when our understanding of nuclear fusion was still pretty spotty—the neutron wasn't discovered until 1932, after all. The notion that nuclear processes powered the stars was just starting to catch on. Only white dwarves were taken seriously as possible stellar end states. In fact, Arthur Eddington, had played such an important role in confirming general relativity, developed a theory in which all stars end their lives as white dwarves.

The Chandrasekhar limit is named for an extraordinary man, Subrahmanyan Chandrasekhar (1910–95). Born in what was then British India, he was such an outstanding student that in 1930 he was awarded a fellowship to pursue his graduate studies at Cambridge University. On the long sea voyage to England, Chandra, as he is known in the scientific community today, worked on the theory of white dwarves. He realized that to exert the kind of pressure that would be needed to prevent gravitational collapse, electrons in the star would have to be moving at an appreciable fraction of the speed of light. This meant that their motion would have to be described by the rules of special relativity and not, as was customary at the time, by ordinary Newtonian mechanics. When he made the calculations, Chandra found that there was a mass for the star above which the electrons would simply be incapable of preventing a collapse. This mass, later named the Chandrasekhar limit, turns out to be 1.44 times the mass of the Sun. In stars above this mass the electrons would simply be incapable of countering gravity. Since there are many stars out there with masses higher than this limit, an unexpected consequence of Chandra's calculation was the statement that there must be other end states for stellar evolution than white dwarves, and indeed we've just pointed out two of them, neutron stars and black holes.

Astronomers at Cambridge didn't pay much attention to Chandra's result, and he went on to obtain a conventional doctoral degree in 1933. There is a folktale in the scientific community that during his thesis defense his examiners, Eddington and astronomer Edward Milne, spent the entire time arguing with each other and ignoring the person they were supposed to be questioning. In any case, Chandra was awarded a fellowship that allowed him to stay on at Cambridge.

What followed was a series of events in which Eddington's behavior is difficult to understand. The kindest word we can come up with is "reprehensible." Eddington appeared to encourage Chandra in his work, dropping by his office and having lunch together, the kind of things that a senior scientist can do that can seem very important to a young scientist. Then, at a meeting of the Royal Astronomical Society in 1935, Eddington arranged to be on the program right after Chandra's presentation and spent all his time belittling it. Using the kind of debating skills common among Oxbridge dons, he continued these sarcastic assaults for years, even after prominent theoretical physicists who looked at the arguments agreed with Chandra. Many science historians have argued that the hiatus in black hole research from the 1930s to the 1960s was due at least in part to people's reluctance to suffer the kind of bullying Eddington was displaying, and hence to avoid the risk of being subjected to his withering assaults.

Reading about this historical incident makes one wonder why Eddington behaved the way he did. One explanation is that Eddington was deeply involved in developing a fundamental theory that would unite all of the sciences, and he was willing to attack anyone who came up with an idea that challenged it, such as Chandra's notion of stellar end states other than white dwarves.

Well, in science as in life, what goes around comes around. Eddington's grand theory has sunk into oblivion. His grand claims to be able to calculate the fundamental constants of nature, such as the charge on the electron, from first principles are now seen as a kind of weird numerology. Meanwhile, Chandra's star rose high. In 1983 he shared the Nobel Prize in physics, after having been honored by many other major awards. When NASA launched the largest orbiting X-ray telescope in 1999, it was named for him. Chandra is still up there, exploring an important part of the universe.

## WHY DID SCIENTISTS REJECT THE EXISTENCE OF BLACK HOLES?

With 20-20 hindsight, it's easy to question the resistance to the notion of black holes in the early and mid-twentieth century. We have to remember, though, that the stellar life cycle outlined earlier, presented in every introductory astronomy course today, was in the process of being developed during this period. It wasn't until 1938, just before the pause in research triggered by World War II, that the German American physicist Hans Bethe (1906–2005) worked out the theory of nuclear fusion that powered stars like the Sun. It wasn't until the mid-twentieth century that astronomers had a firm grasp on the processes that make a star shine, much less any notion of the possible ways that stars could die. It was also in 1938 that physicists J. Robert Oppenheimer and Hartland Snyder, working at Berkeley, far from Eddington's baleful influence, produced what is essentially the modern theory of Schwarzschild black holes, a development that was quickly eclipsed by the start of World War II.

Because of the general lack of knowledge about the internal working of stars, it's not surprising that astronomers were unwilling to accept something as bizarre and counterintuitive as a black hole. Their arguments were not just blind prejudice, however. Without denying that black holes were predicted by Schwarzschild's solution to the Einstein equations, they tended to concentrate their argument on the process by which a star might collapse into one of these strange objects. These arguments, resting as they did on the well-known behavior of fluids and gases, had to be taken seriously.

One of Einstein's arguments, for example, was centered on the properties of something known as angular momentum. The easiest way to grasp this property is to picture an ice skater going into a spin. At the start the skater's arms are extended, and she turns relatively slowly. As she pulls her arms in, however, she spins faster and faster. Einstein argued that like the fast-spinning skater, a star in the process of gravitational collapse will be pulling in its arms and spinning faster and faster. Eventually, Einstein argued, the spin would be so fast that centrifugal forces would simply tear the star apart long before it collapsed to the size of the Schwarzschild radius.

Similarly, Jim's Stanford professor based his arguments on the details of the collapse of a fluid sphere. In order to have a Schwarzschild-type collapse, he argued, all of the different segments of the massive sphere would have to arrive at the condensation point at the same time. Given the turbulent nature of the collapse, he argued that it would be extremely unlikely that this would happen. Instead, different chunks would arrive at the condensation point at slightly different times, and there would never be enough matter at the point at any given time to trigger a gravitational collapse.

There is nothing wrong with raising these kinds of objections to the existence of black holes. It's simply an example of the way that the scientific process works. Whenever a new idea surfaces scientists are expected to raise objections, and every new scientific idea has to go through a testing period. As it happens, detailed mathematical arguments can answer these sorts of objections. Theorists are good at finding ways around these kinds of problems.

Our sense, however, is that it wasn't the problems of fluid flow that kept scientists away from black holes. It was the total repugnance of the notion of the existence of a singularity. Even though the Schwarzschild singularity could never be observed, since nothing can carry information out of a black hole, the very fact that it might be lurking inside a black hole bothered physicists. It is far easier to dispense with black holes entirely than to live with the possibility of a singularity.

The ultimate test in a situation like this would be the discovery of black holes in the sky. It becomes difficult to maintain that something can't form or exist when you have an example of it sitting in front of you. In the 1960s this is exactly the situation in which astronomers found themselves. It's not so much that they went out to find proof of the existence of black holes as that they kept stumbling over things that could be explained only by their workings.

# 5

# THE FIRST STELLAR
# BLACK HOLE

## X marks the spot

*Traditional treasure map*

Think for a moment about what is going on as you read these words. Somewhere something is emitting light: it might be the Sun, an overhead lamp, or a computer screen. If you are reading a book, the light is hitting the page, where it is absorbed by the dark type and reflected by the white paper. That patterned light then travels to your eye, where a complex system of receivers and neurons allows your brain to decode the message the light is bringing in.

Let's look at just one part of this process: the transmission of light, first to the page and then to your eye. As commonplace as it sounds, this process reveals a fundamental truth about the universe that has made astronomy possible through the ages: Earth's atmosphere is transparent to visible light.

You have had many experiences that illustrate this fact. Perhaps you saw the lights of a distant city when you were in an airplane at night, or perhaps you looked up in the dark and saw a full moon or a star. In that last example, note that the light you are seeing traveled vast distances through the vacuum of space before coming through many miles of Earth's atmosphere to your eye. This is the key point: if light were unable to pass through the atmosphere, we could never see the stars and there would be no astronomy.

In chapter 1 we pointed out that visible light is only a small part of the spectrum of electromagnetic waves, a spectrum that stretches from radio waves to X-rays and gamma rays. All the colors of the rainbow, as we have noted, are associated with waves that have wavelengths stretching from about eight thousand atomic diameters for red light to about four thousand atomic diameters for blue light. The reason this small band of wavelengths seems to be so important to us is that we are equipped with sensors especially adopted to detecting electromagnetic radiation in this range, our eyes.

But what about the rest of the electromagnetic spectrum? Again, we can look at everyday experience to guide us to an answer to this question. Think of driving in a car and turning on the radio. The fact that you can hear something means that a signal is traveling from the radio station to your car, and this in turn means that the atmosphere must be as transparent to radio waves as it is to visible light. In fact, we often refer to two windows in the atmosphere, one allowing the transmission of visible light and the other allowing the transmission of radio waves. As it happens, these are the only two windows in the atmosphere. All other types of radiation are absorbed when they try to move through the air.

This explains why until recently there were only two kinds of astronomy possible: traditional optical astronomy, which probes the sky with light gathering telescopes, and radio astronomy, which is characterized by large dish type receivers. We'll talk about this branch of astronomy in more detail when we describe the Event Horizon Telescope in chapter 8. The point, however, is that whatever type of electromagnetic radiation is being emitted by objects in the sky, only visible light and radio waves will make it through that last few miles of their journey and arrive at the surface of the earth, where most of our telescopes are located.

A word of explanation: different kinds of events in the sky generate radiation in different parts of the electromagnetic spectrum. The basic laws of physics tell us that every object with a temperature above absolute zero emits electromagnetic radiation. Right now, for example, your body is giving off infrared radiation because it is at a temperature of about 97.5°F (36.4°C). You don't notice this process because at the same time you are also absorbing radiation given off by objects in your surroundings, which are normally at a lower temperature.

Infrared radiation given off by cool clouds of dust in the galactic center posed serious problems for the astronomers who discovered the first supermassive black hole. But here we want to talk about another kind of astronomical event, one violent and explosive enough to generate X-rays. Radiation from such events can travel billions of light-years through the vacuum of space but can't get through those last few miles of Earth's atmosphere to be registered in our telescopes.

It's obvious that the only way to get around this problem is to put our X-ray detectors above the atmosphere. This is easy enough today, when private corporations are launching rockets all the time, but think of what it must have been like in the early and mid-twentieth century, when interest in exploring the X-ray sky was just starting. The only tools available to put detectors above most of the atmosphere were high-altitude balloons and primitive rockets. Most of these didn't get above the atmosphere, or if they did, they didn't stay aloft long enough to do serious surveys of the X-ray sky. Nevertheless, even in those early days serious studies of X-rays and ultraviolet radiation from the Sun were carried out, setting the stage for what was to follow.

World War II changed the sluggish pace of rocketry development. German scientists, under the leadership of Wernher von Braun (1912–77) began developing advanced rockets, culminating in the development of the V-2. (The "V" came from the German *Vergeltungswaffe*, which translates to "vengeance weapon.") The first long-range rocket, it traveled at 3,500 miles per hour (5,633 kph), had a range of two hundred miles (322 km), and carried an explosive payload of more than one ton. It was also the first rocket capable of getting above the atmosphere, having ascended more than fifty miles (80.5 km) on a test flight. More than three thousand of these rockets were launched in the last year of the war, though they had little effect on the final outcome.

German technological advances in fields like rocketry and aeronautics quickly came to the attention of Allied military authorities and led to one of the strangest scientific stories ever told. Operation Paperclip was originally designed to bring captured German scientists to the United States to help in the continuing war with Japan. It quickly morphed into a program to bring in as many German scientists as possible to give the West an advantage in the burgeoning Cold War. The United States had an unexpected advantage in this endeavor, because many German scientists had sought refuge there when the Nazis came to power. Now these men could be recruited to seek out their former colleagues. It came as a surprise to many young scientists in later years to hear tales of derring-do from their staid older colleagues who had participated in Operation Paperclip.

Concerned about the advance of the Red Army from the east, von Braun led his entire team of scientist, engineers, and technicians west to surrender to American forces. The group was eventually sent to what was then called White Sands Proving Ground in New Mexico, with 177 scientists, engineers, and technicians accompanied by three hundred railroad cars full of captured V-2 rockets and parts. Their mission was to train military personnel in rocketry and set up an American rocket program. The American intelligence apparatus, driven by the demands of the Cold War, decided to ignore the wartime record of many people who were now important assets that could be deployed against the Soviet Union. In many cases documents were classified "secret" to shield potential war criminals from trial.

The early V-2 days at White Sands had mixed results. Rockets kept crashing and exploding. There was even a mild international incident when a V-2 went off course and crashed in a cemetery near Juárez, Mexico. Eventually things got under control, and in 1948 a V-2 took off and carried an X-ray detector high above the atmosphere to measure X-rays emitted by the Sun. It wasn't until 1959 that the Soviet Union was able to launch one of their V-2s, leading to the standing joke among American military personnel that "Our Germans are better than their Germans."

Improved rockets were quickly developed, and by 1952 American-made Aerobee rockets were carrying Geiger counters into space to detect X-rays. At this point, von Braun's team was sent to Redstone Arsenal in Alabama, where it was eventually involved in building the Saturn rocket, which carried Neil Armstrong and Buzz Aldrin to the Moon—arguably one of the greatest achievements of the human race.

Probing the sky for X-ray sources other than the Sun became a major scientific program in the 1960s. In 1964 a sounding rocket, that is, one that did not travel into space, picked up an X-ray signal from a system about 7,200 light-years from Earth in the constellation Cygnus, the Swan. This system was subsequently named CYG X-1, following scientific nomenclature by which the sky is broken up into eighty-eight constellations, more or less modeled after the ancient groupings of stars; the first part of an X-ray source's name gives the constellation in which it is located, while the number specifies the order in which the sources in a given constellation was discovered. Thus, CYG X-1 was the first X-ray source discovered in the constellation Cygnus.

What astronomers found when they examined the place in the sky from which the X-rays were coming was a single supergiant blue star. As such, that was nothing special. Watching that star, however, revealed that the light it emitted alternately red- and blue-shifted. In other words, over a period of about five days, the star was either moving toward us or away from us. The only way to explain this behavior was to say that the star was in orbit around a massive object. That object was the source of these X-rays, but it couldn't be seen in our telescopes. Furthermore, data from the orbital motion of the visible star indicated that the mass of the companion was well above the analogue of the Chandrasekhar limit (see chapter 4) for neutron stars. At the time, astronomers calculated that the companion had to be at least six times as massive as the Sun—a value that was later raised to twenty-one times as massive as the Sun. Either of these masses would be high enough to show that the companion object couldn't be a neutron star, and that left only one option: it must be a stellar black hole.

The picture of the Cygnus X-1 system that has developed over the years reflects the kind of complexity we find when we look at the heavens in detail. In the beginning, the system consisted of two massive stars circling each other. One star went through the kind of life cycle we talked about in chapter 4, evolving into a black hole. The intense gravitation field of the black hole began pulling material off the supergiant star, which was still in its main sequence stage. This material started to fall into the black hole, forming what is called an accretion disk, a structure of great importance to our discussion of all types of black holes.

The main way that a black hole influences its surroundings is through its gravitational field. This field will attract material in the black hole's vicinity. The black hole isn't like a vacuum cleaner, however, just scooping up anything in its neighborhood. To see this, think for a moment about a rock floating around in the neighborhood of a black hole. Under the influence of the black hole's gravity, it may indeed fall into the hole—*provided* it is moving straight toward the black hole at the start, which is unlikely. More likely, the rock will have a velocity that is tangential to the hole, just as the planets in our solar system move in orbits around the Sun instead of falling straight in. Thus, as the black hole pulls in nearby material, that material is most likely to go into some kind of orbit instead of falling straight in.

This means that if there were only one rock in the neighborhood of the black hole, it would most likely go into a planetlike orbit. In the real world, however, there will be all sorts of stuff being pulled in, not just a single rock. This means that as the black hole moves through the interstellar medium, the material being pulled in will start to accumulate in orbits, beginning to form what we have called an accretion disk. At this point, processes more complex than simple Newtonian gravitation begin to show up. Bits of material in the disk begin to collide with each other, and processes such as turbulence and the formation of eddies begin to show up. Collisions begin to raise the temperature of the material in the disk. As the temperature of the disk goes up, the electromagnetic radiation that is emitted works its way into the range of X-rays.

CYG X-1 started out as a normal double star system. There's nothing unusual there; probably half the stars you see in the sky are in multiple star systems. One of the stars was big, and in a short time, astronomically speaking, it ran through its hydrogen and wound up as a black hole. It started pulling material off its partner and began forming an accretion disk. No surprise, then, that when astronomers began examining the region that seemed to be the source of X-rays, all they saw was a single star that seemed to be jerked around by an unseen partner. It was material from that companion star, not yet through its life cycle, that turned the system into a major X-ray source. The temperature in the disk climbed to millions of degrees. At that temperature matter starts giving off X-rays, and this is what attracted the attention of those scientists at White Sands so many years ago.

The discovery and interpretation of Cygnus X-1 settled forever the debate on the existence of black holes. They were real, and here was one of them right on our doorstep. It is important to realize, however, that we do not "see" the black hole in CYG X-1—what we see are the effects that the black hole has on matter being pulled away from the companion star. Even that will be gone in a short time (astronomically speaking) once that companion star has gone through its life cycle and become a neutron star or a black hole. In that case, there will be no way to pull matter from the companion and the X-rays will vanish.

This illustrates an important point about black holes. If they result from the collapse of solitary stars, there is no easy way that they can be detected. It is estimated that there may be a million or more such solitary black holes wandering around the Milky Way, unknown and undetected.

We have to point out one more issue illustrated by the discovery of CYG X-1. We have obviously found an object that looks like the black holes predicted by the Einstein equations. We have not, however, proved that this is the predicted object. This may seem like a minor point, but it's something we may want to keep in mind. We have established that there are objects out there that have all the characteristics of the predicted black holes. They probably *are* the predicted black holes. Nevertheless, we should keep in mind that this is an assumption, not a statement that has been proven.

Though historically important in establishing the existence of black holes, stellar black holes like CYG X-1 have had little effect on the evolution of the universe. In what follows we will be primarily concerned with what are called supermassive black holes.

# 6

## SEARCHING FOR
## THE CENTER

———

In enterprise of martial kind
When there was any fighting
He led his regiment from behind
He found it less exciting

*W. S. Gilbert and Arthur Sullivan,* The Gondoliers

**W**illiam Herschel (1738–1822) was second oboe in a military band in his native Hanover, hardly what you might expect for a man who was to begin a revolution in the way human beings thought about their place in the universe. After he was involved in a major defeat for the Hanoverian forces, he left the army (some historians have labeled him a deserter) and emigrated to England. There he eventually took a position as organist in a church in Bath and, with his sister Caroline, the two began careers that we would say today were typical of serious amateur astronomers.

In those days, before the introduction of street lighting, the view from a backyard in a city like Bath would be equivalent to what you would see in a remote area far away from any city today. In addition to being a musician, Herschel was a master craftsman and telescope builder, who in later life would supplement his income by selling telescopes to other astronomers, who regarded Herschel telescopes in roughly the same way a violinist values a Stradivarius today. We have complaints in Caroline's diary of her brother moving various machines into rooms in their house, as well as a hilarious account from Herschel himself of an incident when molten metal escaped onto a floor. The flagstones in the floor exploded, with pieces bouncing off the ceiling and making the observers, who included a knight and a member of the Royal Society, flee for their lives. The incident occurred because in those days telescope mirrors were made of polished metal, not glass.

Herschel's most famous discovery occurred in 1781, when he discovered the planet Uranus, the first planet not visible to the naked eye. He first proposed naming it for King George III of England. This never caught on with astronomers, but it did get Herschel a royal pension that allowed him to pursue astronomy full time. From our point of view, however, the most important work the Herschels did was to attempt to create a detailed map of the Milky Way.

In fact, they spent more than fifteen years looking at more than six hundred directions in the sky, painstakingly counting every single star they could see with their telescope, with a forty-eight-inch mirror the largest in the world at the time. By assuming that each star had the same intrinsic brightness—that is, that brighter stars were closer and dimmer stars farther away—they calculated distances to each star, creating the first detailed map of the whole sky. What they reported was that the galaxy was basically a disk of stars with the Sun roughly in the center. Generations of astronomers followed their study, armed with bigger and bigger telescopes, culminating with a more detailed map based on star counting put forward by Jacobus Kapteyn in publications issued between 1896 and 1900.

But there were two fatal flaws in this approach to mapping the Milky Way. For one thing, astronomers at the time did not know how to measure accurate distances to distant stars, particularly those farthest away. In addition, they thought that space was empty except for the stars. They did not realize the existence of one ubiquitous component of space: interstellar dust. Their view of the distant parts of the galaxy was being blocked by that dust, and as a consequence they could not see either the center or the edge of the Milky Way. They were like someone driving in fog, to whom all that is visible is a sphere around the car, a sphere whose radius is determined by the density of the fog. In the same way, the Sun-centered sphere the Herschels reported was a product of interstellar dust playing the role of the fog in our analogy.

## THE DUSTY UNIVERSE

The unsuspected presence of interstellar dust played such an important role in the process of mapping out our galaxy that we should take a few minutes to think about its role in galactic astronomy. As it happens, astronomers have known about the effects of interstellar dust for almost as they have known about stars. You can actually see these effects with your naked eye. If you can see the Milky Way, you will notice a dark streak running down its middle. This dark spot is typical of what astronomers saw when they surveyed the galaxy with their telescopes—the dark spots seemed to be everywhere.

There are, of course, two ways to explain the presence of such dark areas. One is to say that they are, for whatever reasons, regions where there are very few stars. William Herschel believed that this was the reason for the dark areas, and he called them "holes in the heavens." He believed that if you looked through them you would see the true structure of the universe. The other possible explanation is that the dark regions are filled with a substance that absorbs light so that whatever light is emitted by objects behind them is absorbed and never reaches our telescopes.

It wasn't until 1930 that the Swiss American astronomer Robert Trumpler resolved this issue. He looked at groupings of stars called open clusters (typically consisting of a few hundred young stars), using Cepheid variables to get the distance to each cluster. (We'll describe Cepheid variables in the next section.) He then calculated how much light we should receive from other stars in the cluster, given their distance from us. He found that less light was coming into our telescopes than we would have received had the light traveled from the cluster without interference. The only conclusion possible is that something is absorbing the light while it is traveling toward us. Subsequent studies establish that that "something" are ubiquitous dust grains scattered between the stars.

In fact, we know that interstellar dust is composed of small grains about the same size as the wavelength of visible light. These grains can have a core made of carbon bearing minerals (called "soot") or silicon minerals (called "sand"). When light encounters these sorts of grains, one of three things can happen:

- It can be absorbed. In this case, as Trumpler found, the star will appear to be fainter than it would be in the absence of dust. Note that in the process of absorbing light the grain will heat up and radiate in the infrared range, an effect we will discuss in the next chapter. The reduced intensity is usually referred to as "extinction."
- It can be scattered. The laws of physics tell us that short wavelengths (blue light) will be scattered more than long wavelengths (red light). Scattered light doesn't reach our telescopes, so we see light coming in with the blue and green wavelengths removed—a fact that makes

the light appear red. Note that we see this same effect when we see a red sunset. In that case the light is being scattered by molecules in Earth's atmosphere. We see the same effect when we look at a blue sky: in this case we're seeing the blue light that has been scattered from the beam. Astronomers call this effect "reddening," but it's not that red is being added to the incoming light—it's that blue and green are being removed. One author even suggested referring to the astronomical effect as "de-bluing."

- It can be reflected. If there is a dense cloud near a bright star, we may see a fuzzy halo of light being reflected from dust grains in the cloud. Note that this is not light that originates in the grains; the grains are merely a mirror that redirects the light to us. The constellation of the Pleiades is often given as an example of this effect.

Until recently, we had to rely on indirect (but strong) evidence for our understanding of interstellar dust. In 1999, however, NASA launched the Stardust space probe. The probe's primary mission concerned a visit to comet Wild (pronounced Vilt)-2, but as a secondary goal it was to collect some interstellar dust. For this task it was equipped with two retractable tennis racquet sized collectors made of an aerogel capable of slowing down and capturing passing interstellar grains. In 2006 the spacecraft returned its samples—cometary plus interstellar—when a special capsule landed in the Utah desert and was recovered. Thanks to Stardust, we now have sample of what's out there between the stars.

## HENRIETTA LEAVITT AND THE DISTANCE SCALE

How we came to understand the true size and shape of our galaxy, and that our galaxy is just one among possibly a trillion galaxies that extend out to enormous distances is one of the most exciting journeys in human history.

The journey to unlock the true nature of galaxies began with a remarkable discovery made by Henrietta Swan Leavitt (1868–1921) in 1912. She was brought to Harvard University as one of a group of people (mostly women) who carried out the detailed mathematical calculations needed by astronomers. Leavitt conducted a tedious project that required her to analyze the brightness of stars captured on photographic plates in the Harvard collection. She noticed that a group of variable stars, first discovered in the Southern Hemisphere in the constellation Cephus and known as Cepheid variables, had an unusual characteristic: the brightness of the stars changed on a regular cyclical basis. When she looked at Cepheid variable stars that were close enough to Earth so that their distance could be easily determined by triangulation, she found that the time required for the brightening-dimming cycle, a process that would take weeks or months to complete, seemed to depend on the brightness of the star. Watch the star go through its cycle, in other words, and you can calculate its brightness, and by comparing that to how much light we receive from the star we can calculate its distance from us. The Cepheid variable became what is known as a "standard candle," the first of several we will talk about in later chapters.

The Cepheid variable can be said to be the tool that launched modern astronomy. Yet because she was a woman, hired as a "computer," Leavitt was excluded from the follow-up work that transformed our understanding of the cosmos, and, sadly, she died without being able to participate fully in the science that her work launched.

## HUMANITY'S PLACE IN THE UNIVERSE

One of the important tasks of the sciences, and particularly the science of astronomy, is to supply human beings with a sense of where they fit into the universe. For millennia, for example, the world's greatest astronomers taught that Earth was the unmoving center of creation.

Besides being obvious to our senses, this notion fit in well with the way people saw the natural world. Dating all the way back to Aristotle, the *scala naturae* (ladder of nature) gave a simple and flattering view of the natural world. Often depicted as a layered pyramid, the bottom layer was all the nonliving material in the world. Moving upward, we find layers of increasing "perfection," from plants to animals until, just below God and the angels, we find human beings. Given our exalted place in this scheme, it was perfectly reasonable to suppose that our home, Earth, should have a special place in creation. Even when Nicolaus Copernicus showed that the Sun, and not Earth, was the center of the solar system, we could take solace from the fact that our Sun was at the center of the galaxy. No wonder the Herschels' result was so readily accepted in a time when most astronomers believed that our own galaxy was the only one around.

Who would have believed that a lone woman, toiling in obscurity measuring smudges on photographic plates taken by generations of astronomers would give her colleagues the tool they needed to develop our modern picture of the importance of our home planet?

Yet that's exactly what happened.

The reason is simple. If we find a collection of stars that includes one or more Cepheid variables, Leavitt's technique can be used to find the distance to that collection of stars. The first person to make use of this process was the American astronomer Harlow Shapley (1885–1972), the longtime director of the Harvard College Observatory. Shapley studied large clusters of stars, known as globular clusters orbiting in or near the constellation Sagittarius, the Archer. Shapley suspected that his results meant that the center of the galaxy was not near the Sun but somewhere in Sagittarius. Armed with Leavitt's discovery of pulsating stars, he was able to measure the distance to these clusters more accurately than his predecessors. He not only saw that the stars were orbiting a center far from the Sun in Sagittarius, but he also found that the Milky Way galaxy was about ten times larger than anyone had believed up to that time.

There was one other odd attempt to locate the galactic center behind all that dust. It occurred in the early days of World War II. After the Japanese attack on Pearl Harbor, people were worried that here might be bombing attacks on West Coast cities. For a while, governments instituted blackouts: lights had to be turned off and windows covered up at night. Astronomers at Mount Palomar Observatory took advantage of the darkness to confirm Shapley's measurements of the globular clusters and verify that they seemed to be in orbit around something that could be the galactic center.

Although Shapley and his contemporaries suspected the center of the galaxy resided somewhere in Sagittarius, the precise location was not known. However hard they tried, telescopes at the time just could not see anything at the core of the globular cluster orbits. The problem was, of course, dust—lots of dust. They just didn't know about its existence. And this dust prevented any view of the heart of the nuclear regions of the galaxy with optical telescopes.

Shapley was the head of a large East Coast research institute, but he never lost his midwestern sense of humor. (He was born in Missouri.) He delighted in telling reporters that he had become an astronomer because when he entered college he began leafing through the college course catalogue to find a major. "I couldn't spell archeology," he would say, "and astronomy was next in line."

Today, astronomers have a pretty good idea of the place of Earth in the galaxy. Here it is: right now, you and everything that surrounds you, the entire Earth and the Sun, are moving at a speed of 125 miles (201 km) per second around the center of the Milky Way, roughly 26,000 light-years away. It takes 220 million years to make one trip around the center, and the Sun and its planets have done so nineteen times in cosmic history. What lies in the center of our galaxy, however, remained a secret until quite recently.

Before we resume our search for the galactic enter, however, there is one more result of Leavitt's work that we need to discuss. In 1924 the American astronomer Edwin Hubble (1889–1953) studied Cepheid variables in the Andromeda nebula. This is a fuzzy blob observable with the naked eye, thought at the time to be a cloud in our own star system. What Hubble found shook the astronomical community: Andromeda was 2.5 million light-years away. It certainly was not in our star system, but it was an entirely separate star system a vast distance away. The Andromeda nebula, in other words, wasn't a simple nearby cloud, but another galaxy like the Milky Way.

We've all gotten used to the idea that the Milky Way is just one of billions of galaxies in the observable universe, but this was a hard pill for many scientists to swallow at the time. Before Hubble's discovery, in fact, there was a lively debate among astronomers about whether things like Andromeda were inside the Milky Way or other "Island Universes," to use a term popular at the time. In fact, in 1920, the National Academy of Sciences sponsored a debate between Shapley and astronomer Heber Curtis on this very issue, with Shapley arguing that they were inside the Milky Way. Hubble's discovery settled this issue a few years later.

There is, of course, one more important result of Hubble's work. Leavitt's technique gave him the distance to what he now knew were distant galaxies. By measuring the wavelength of light emitted by atoms in those galaxies, we could calculate how fast they were moving. He found that the distant galaxies are moving away from us, and the farther away they are the faster they are moving. The universe, in other works, is expanding. This is the origin of the Big Bang theory.

So those photographic plates at Harvard gave us the tool we needed to show that the Sun is not the center of the galaxy. The universe is actually a huge set of galaxies like the Milky Way, and the universe is expanding. That's a whole lot of information to get from some smudges on a set of glass plates.

## SEEING THROUGH THE DUST

By the 1930s, then, we knew that stars in the universe were gathered into galaxies and that we lived in a typical galaxy we called the Milky Way. Here's where that name comes from: on a moonless night far from city lights during certain times of the year, you will see a pale streak of light that extends from horizon to horizon. Named *galaxies kuklos*, or "milky circle" by the Greeks, and later *via lactea*, the "milky road," by the Romans, this wispy band of light was a mystery until Galileo pointed his rudimentary telescope to the night sky in 1610. To his amazement, he was able to see that this band in the sky that we call the Milky Way consisted of innumerable stars that all blended together and were too faint to see with the naked eye. Today we know that our galaxy is a thin disk of stars, and the Milky Way is what the disk looks like to someone on the inside looking out. Once we understand that we are located inside of a galaxy, our next task is to understand its structure. And, of course, the first task is to find and analyze the center.

And here we run into the problem of the dust. Because the dust absorbs visible light, an observer on Earth, like the driver in a fog, can see only what is inside his or her "sphere of visibility." It makes no difference how big and powerful our telescopes are; the dust absorbs the light before it gets to us. We had to find a new way of observing if we wanted to see the center of the galaxy.

Oddly enough, that new way of observing grew out of a very practical problem faced by people in the nineteenth century, communication between Europe and the burgeoning industrial giant in North America. It took sailing ships at least ten days to make the trip. In 1858 the first transatlantic cable was connected between England and

America. The first message, a congratulatory message from Queen Victoria to President James Buchanan, was ninety-eight words long and took sixteen and a half hours to transmit. It wasn't until 1901 that Guglielmo Marconi (1874–1937), generally regarded as the man who developed radio communication, actually sent a transatlantic radio message from England to Cape Cod.

That call contained a lot of background noise, which was a real hindrance to the emerging technology that would soon transform the world. Eventually this got to be such a problem that Bell Telephone Laboratories, a telecommunications company based in New Jersey, hired a young physicist named Karl Jansky (1905–50) to try to find the origin of the background noise.

Jansky's father was born into an émigré community from what is now the Czech Republic and was a professor of engineering at the University of Oklahoma, introducing Janksy to science at an early age. After receiving a bachelor's degree in physics from the University of Wisconsin, he was an ideal choice to start a research program at Bell Labs designed to improve wireless transmission.

At Bell Labs he built a moveable radio receiver to see where the static came from. The receiver was an ungainly thing, more than a hundred feet (30.5 m) long, twenty feet (6.1 m) tall, and mounted on a set of tires scavenged from an old Model T sedan. Jansky's colleagues dubbed the apparatus, constructed on an abandoned potato farm, "Jansky's Merry Go Round." Ungainly or not, it worked, and Jansky was soon able to show that some subclasses of static were generated by terrestrial thunderstorms. No matter what he did, however, he couldn't locate the source of a constant hiss that the antenna kept picking up.

As time went by and the direction to the source of the hiss shifted in the sky. Jansky realized that the source had to be extraterrestrial—that is, it came to Earth from the stars. He found that the signal peaked when the antenna was pointed at the constellation Sagittarius—the very spot that Shapley had suggested as the galactic center.

We would love to report that the scientific community immediately used Jansky's techniques to examine the heart of the Milky Way, but that's not what happened. After all, here was this kid with nothing but a bachelor's degree and absolutely no credentials in astronomy talking about measurements that made no sense to seasoned astronomers. When Jansky presented his work to a small group of astronomers in 1933 few of the professionals paid much attention, although on May 5, the *New York Times* gave quite a bit of space to the "extraterrestrial signals." Nevertheless, the key point of Jansky's work was that radio signals, unlike visible light, could come through the interstellar dust from distant sources. This came to be the founding idea behind the field now known as radio astronomy.

Unfortunately, Jansky's epic discoveries came in the middle of the Great Depression. Convincing Bell Labs to put serious money behind a new research project of uncertain profitability was difficult, to say the least. Jansky went back to Wisconsin to earn a master's degree (with a thesis on "Star Noise") and spent the rest of his career at Bell Labs working on improving radio transmissions until his untimely death at age forty-four.

Perhaps because of the Great Depression, the notion of using radio waves to probe the heavens was dormant for a while after Jansky's discovery. Today, as noted, he is recognized as the founding father of radio astronomy. In the words of a later scientist, John Krauss: "In 1930 essentially all that we knew about the heavens had come from what we could see or photograph. Karl Jansky changed all that. A universe of radio sounds to which mankind had been deaf since time immemorial now suddenly burst forth in full chorus."

# 7

## THE MONSTER AT THE
## HEART OF THE GALAXY

---

### Never trust your advisor!

*Advice of Nobel Laureate Charles Townes to graduate students*

I f you know that there's something with unknown mass and size somewhere out there, with all kinds of stuff in orbit, how do you proceed? One way of monitoring movement was provided by the Czech-Austrian mathematician Christian Doppler (1803–53). He looked at a fundamental property of waves, and as a result the so-called Doppler effect is used today to monitor everything from weather patterns to blood flow. Although the Doppler effect is present in any kind of wave, it is perhaps easiest to visualize by picturing waves on water.

Imagine that you have a machine that causes a paddle to strike the water in a pond at regular intervals. The machine will send out concentric ripples, and the distance between crests is known as the wavelength of the outgoing waves. Let's start by assuming that both you and the ripple machine are stationary. In this case, the wavelength you measure when the ripples get to you will be the same as the wavelength of the wave created by the machine.

Now suppose the machine is moving toward you. The distance between the crests of the crest of one ripple and the next will be shortened by the distance the machine has moved between the emission of successive crests. Consequently, you will see the wavelength of the wave that arrives at your position shortened. A similar argument should convince you that if the ripple machine is moving away from you, you will see a longer wavelength.

This simple argument can be applied to any wave. For sound, a longer wavelength corresponds to a lower pitch, while a shorter wavelength corresponds to a higher one. This explains the sound you hear when race cars go by you on a track. For light, on the other hand, long wavelengths correspond to red colors, while short wavelengths correspond to blue. This insight is the origin of the "red shift" that Hubble used to establish the expansion of the universe (see chapter 6). If you know the wavelength of the wave being produced by the machine when it is not moving, you can use the Doppler effect, along with the wavelength you actually measure, to determine the velocity of the source, and this is where the Doppler effect begins to have an important impact in astronomy.

The main questions we want to answer when probing the galactic center are these: What is the mass of the central body, and what is its size? Depending on the kind of black hole we're looking at, finding the mass can be a complex process. We can think of the various techniques as being divide into "direct" and indirect categories. The direct categories look for emissions from material that is under the direct influence of the gravitational field of the black hole. These objects can be stars or gas clouds that can absorb or emit radiation. We can often watch these stars and gas clouds move. It turns out, in fact, that they move at unbelievable speeds if they are near the central black hole—10 million miles an hour (16 million kph) is not unusual—and if we observe long enough we can map out entire orbits. Once we have this, it is a relatively simple task to use standard Newtonian mechanics to get the mass of the black hole.

Analysis using the indirect methods is more complicated. Basically, we have to use some correlation between the properties of the host galaxy and the mass of the central black hole that has been established for other galaxies (by use of the direct method, for example) and assume that the same correlation exists for the galaxy we're examining. If there is no hot gas near the black hole, as is the case for most galaxies, we may be able to pick up radiation from groups of stars, even if we can't see them individually.

The fact of the matter is that the atoms of various chemical elements emit well-known collections of electromagnetic waves as electrons shift around between atomic orbits. These waves are typically in the range of visible light, and that causes a problem if we try to use atomic emissions to determine the speed of objects near the galactic center. Even if material near the galactic center is emitting clean electromagnetic signals, the intervening dust guarantees that those signals will be absorbed long before they get to Earth. In addition, the atoms in the accretion disk have typically lost their electrons through collisions, so that the material is actually what physicists call a plasma. Atomic emissions like those described earlier will be seen only in cooler regions farther from the central black hole.

If we look at radio waves that can penetrate the dust, we run into another problem. In general, stars are very weak radio emitters, and this means that our radio telescopes aren't going to be able to see them directly. Radio measurements will most likely detect gas clouds rather than stars, and this means that our interpretation of the data will be complicated by the presence of things like turbulence and eddies, a fact that casts doubt on any calculation we make of the mass of the central object. In any case, given the shortcomings of both optical and radio detectors, we obviously need a new kind of probe to explore the galactic center.

Fortunately, William Herschel started us on the track to that new kind of probe way back in 1800. He wanted to measure the way that energy was divided among the various colors in sunlight, so he set up a system in which sunlight was run through a prism and then placed thermometers in different color bands to track the energy. He was surprised to find that if he put one of his thermometers in what appeared to be a blank space beyond red, the longest wavelength of visible light, the thermometer registered the presence of incoming energy—energy invisible to the human eye. Herschel called what he saw "calorific rays," but we call it infrared radiation today, a term derived from the Latin *infra*, which means "below."

Infrared radiation plays an important role in many fields beside astronomy. The reason for this lies in one of the basic facts of thermodynamics. It is a fact that every object in the universe at a temperature above absolute zero (−273°C) emits electromagnetic radiation, with the wavelength of that radiation depending on the object's temperature. It turns out that objects at around room temperature emit infrared radiation, and this explains why infrared radiation has become so important in our modern technology. It is the basis, among other things, of all night vision devices, which operate because objects at room temperature emit infrared radiation even when they do not emit or reflect visible light.

Throughout the nineteenth and early twentieth century, however, the development of infrared detectors was something of a sideshow, not really regarded as an important endeavor. Research focused on finding materials that would produce an electric current when they absorbed infrared radiation. A few materials, such as lead sulfide (PbS), were found, but nobody took the technology all that seriously.

As often happens with new technologies, the initial developers and users were various national military organizations. During World War II, for example, the German Wehrmacht developed a small infrared telescope that could be mounted on a sniper's rifle. Called the *vampir* (vampire), it was a bulky thing, weighing in at about thirty-three pounds (15 kg) with its battery pack. The device saw limited service on the eastern front in the last days of the war. The elite corps that used this device were known as *Nachtjäger* (night hunters) to emphasize their ability to see at night.

It wasn't until the 1970s that infrared receivers were declassified and made available for nonmilitary use. Judy Pipher of Rochester University developed the first infrared detector array. One of the early arrays, obtained from the military, had the words "Tank Breaker" written on its side. It gave us our first infrared views of the outer planets and other galaxies.

Following the invention of the transistor in 1947, there was an explosion in our ability to use semiconducting materials to make infrared detectors, and it wasn't long before astronomers began to put these new materials to use. In effect, a new window into the heavens opened up, a window that allowed us to explore the infrared sky. As far as the galactic center is concerned, the fact that stars are strong emitters of infrared radiation meant that the weakness of their radio signals would no longer prevent us from measuring their velocity.

Infrared telescopes were the first instruments used to give us a picture of the galactic center. It turns out that the stars there are densely packed—there are about 10 million of them. In addition, there are swirling clouds of gas orbiting the central object. Eric Becklin was an astronomy graduate student at Cal Tech, and one night he was alone at an infrared telescope. He told his advisor, Gerald Neugebauer, that he wanted to look at the galactic center. Becklin had no clear purpose for doing so, and Neugebauer discouraged him, saying there was nothing there to see. Becklin persisted, though, and discovered the supermassive black hole called Sagittarius A, or SagA*. Nobel laureate Charles Townes thought this story interesting enough to include in a public lecture from which the epigraph to this chapter is taken.

From the point of view of astronomers, the ability to detect infrared radiation is a good news/bad news story. The good news is that infrared radiation can cut through the dust clouds between us and the galactic center. The bad news is that infrared radiation can be absorbed in Earth's atmosphere, particularly by the presence of water vapor. In fact, infrared radiation can travel only a limited distance through the atmosphere before it is absorbed; your TV remote, for example, sends a signal that can get to your TV set but not to your neighbor's living room. The problem of absorption can be dealt with by putting infrared telescopes on high mountains, above most of the atmospheric water vapor or, ideally, by putting the detectors in space, above the atmosphere.

There is one issue with regard to infrared telescopes that complicates putting infrared instruments in orbit above Earth's atmosphere. As we pointed out, every object—including an infrared telescope in orbit—emits some sort of electromagnetic radiation. If this is the case, the question becomes this: How do you keep the telescope from seeing itself? How, in other words, do you separate infrared signals from distant object with the same sort of signals being emitted by the telescope itself?

One way of dealing with this issue is to cool the telescope to the point at which it is emitting microwaves rather than infrared radiation. This will indeed allow the telescope not to see itself, but it requires a very low temperature coolant, typically something like liquid helium at a few degrees above absolute zero. The useful lifetime of an infrared satellite, then, is set by the availability of the coolant. A few years in orbit is all we can expect from such an instrument.

In any case, once data on infrared radiation from the galactic center was available, calculations indicated that gravitating central object in the galactic center had mass about 4 *million* times that of the Sun. The attention of the astronomical community turned to the question of what such an object could be. It turned out to be the first supermassive black hole ever detected.

Once the mass of the central object was settled, the next job was to determine its size. Only a combination of these two measurements can establish the presence of a black holes, an object in which a great deal of mass is confined inside a very small radius. In the 1970s and 1980s astronomical measurements of the object in the galactic center were limited by the resolving power of the available telescopes. As we mentioned in chapter 6, measurement of this type showed that whatever this object was, it was no larger than our own solar system. As telescopes got better, this limit shrank a bit. From our point of view, the most interesting thing about this period is the fact that no matter how the measured size of that central object shrank, our estimate of its mass stayed pretty much the same.

There came a time, however, when attempts to determine the size of that object ran into another fundamental problem. Infrared radiation may be able to get through the dust that hides the galactic center from our eyes, but sooner or later it's going to have to come through Earth's atmosphere to our telescopes. Earth's atmosphere is a chaotic thing, with temperature inversions, winds, and difference in pressure and density all over the place. These atmospheric phenomena have the effect of blurring and smearing out the image being carried by electromagnetic waves heading for the surface. Think of it this way: each little tweak in the atmosphere moves the image you are getting to a slightly different position in your telescope. This motion is what smears out the image and limits our ability to measure details of distant objects.

There's no way to stabilize the atmosphere. These atmospheric effects make the stars twinkle, after all. Although people had thought about ways to deal with this issue as early as the 1950s, it wasn't until the 1990s that computers became powerful enough to implement a way to "untwinkle" the stars: adaptive optics.

Here's how it works: a telescope mirror is built in segments, each of which is controlled by a computer guided motor. A wave with known properties coming through the atmosphere is measured (more about this wave in a moment). Whatever the atmosphere is doing at the time this wave passes through will cause distortions in the wave. The computer measures these distortions and adjusts the segments of the mirror to compensate for the measured atmospheric distortions.

The goal of this scheme, then, is to see what the atmosphere has done to a light beam with known properties and then to use that information to adjust the telescope to compensate for the atmospheric disturbances. Obviously this system will work provided we have some way of knowing what the atmosphere has done to that first incoming light beam. In the jargon of astronomers, we will need what is called a "guide star" to tell us what the atmosphere is doing in real time.

There are a number of different types of guide stars. If, for example, there is a bright star near the object you want to examine, you can use the distortion of that star's image to make the adaptive optics system work. Usually, however, we aren't that lucky and have to produce our own guide star. To understand how we can do this, we have to talk a little about Earth's upper atmosphere. The mesosphere is a thin layer fifty to sixty-five miles (80–105 km) above Earth's surface. It is the first part of the atmosphere encountered by incoming meteorites. The heat generated by the friction between the meteorite and the atmosphere starts burning the latter's outer layer. As a consequence, the mesosphere has a layer of burned-off materials, including atoms of sodium.

Sodium vapor is a material that produces a bright yellow glow when heated, for which reason it is widely used for highway illumination. If we shine a laser tuned to that light, it will produce a bright spot when it encounters sodium atoms in the mesosphere. Light from this spot will come down toward the ground, and this means that we can use the spot as an artificial guide star. Most astronomical measurements made with adaptive optics use this system to make their guide stars.

Even applying all of these advanced techniques to observations at the galactic center doesn't necessarily give us an unquestionable explanation of is structure. The radio investigations, for example, convinced most scientists that the massive object at the center of our galaxy was a black hole. Not everyone agreed with this conclusion. Some theorists argued that the object around which everything was rotating was a dense gas cloud.

There is nothing wrong with the fact that people suggested other ways to interpret the data. In fact, coming up with alternative explanation like this is a serious obligation in every field of science. It's simply one way we can keep from fooling ourselves. When it is working at its best, this sort of skepticism drives other scientist to do new experiments and calculations that either support or contradict any new idea.

One installation that makes use of all these advanced techniques is the so-called Very Large Telescope (VLT) operated by the European Southern Observatory in Chile. It's not a single telescope, really, but instead a series of four identical telescopes next to each other. Located in the Atacama Desert, one of the driest and most isolated spots on the planet, it is ideally situated to observe distant galactic centers.

In the case of objects orbiting the galactic center, we are in a situation that you would have if you were standing on the ground watching a merry-go-round. If the merry-go-round is rotating in a counterclockwise direction, then the animals on your right will be going away from you, while the animals on the left will be coming toward you. The velocities of both sides can be obtained from the Doppler shift, so the rotational speed of the object, along with the mass of the central object, can be calculated.

In principle, a calculation can be done for anything rotating around the central object. As we pointed out, however, rotating clouds of gas and dust are subject to all sorts of nongravitational forces. The ideal situation is one in which we have data on stars in orbit around the central mass. Stars aren't going to be pushed around much by gas clouds, so for stars only the gravity exerted by the central object needs to be taken into account.

Starting in the early 1990s Andrea Ghez of the University of California, Los Angeles, and Reinhard Genzel, director of the Max Planck Institute for Extraterrestrial Physics, began monitoring the motion of stars inside the galactic center. These stars are customarily given a designation S0 if they are within a specified distance from the galactic center and a number indicating where in the order of stars they are according to their distance from the center. One star, S0-2 (the second farthest out) has been monitored for the entirety of its sixteen-year orbit. These orbits have established the identity of the central object beyond any doubt: it is a black hole whose mass is 4 million times that of the Sun and whose outer boundary is about as far from its center as Pluto is from the Sun. It took a number of technical advances to establish this fact, and for this work Ghez and Genzel shared part of the Nobel Prize in physics in 2020.

Since its discovery, astronomers have witnessed a number of amazing events in our galaxy's center. Here are a few of them:

- On January 5, 2015, NASA reported observing an X-ray flare four hundred times brighter than usual, a record-breaker, from Sgr A\*. The unusual event may have been caused by the breaking apart of an asteroid falling into the black hole or by the entanglement of magnetic field lines within gas flowing into Sgr A\*.
- On May 13, 2019, astronomers using the Keck Observatory witnessed a sudden brightening of Sgr A\*, which became seventy-five times brighter than usual, suggesting that the supermassive black hole may have encountered another object.
- VLT/Gravity, with its improved resolution in the infrared, uses four different telescopes. It observed star S29, which made its nearest

approach to the black hole in late May 2021. It passed it at a distance of just 8 billion miles (13 billion km), about ninety times the Sun–Earth distance, at the stunning speed of 5,430 miles (8,740 km) per second. No other star has ever been observed to pass that close to, or travel that fast around, the black hole.

From ancestral gazings at the wispy band of light that stretched across the sky to the painstaking work on pulsating stars of a woman working behind closed doors, the first transatlantic phone call, and the gift that meteorites deposited in our atmosphere, the secret hiding in the middle of our galaxy was finally revealed.

# 8

# MONSTERS EVERYWHERE

---

**Seek and ye shall find.**

*Matthew 7:7*

Looking at black holes in the Milky Way, we see that they fall into two categories. On the one hand we have stellar black holes, weighing in at masses up to a few dozen times that of the Sun. We understand that stellar black holes are the natural end state of the life cycle of very large stars. We also understand that we can "see" these types of black holes only through their effects on visible objects like companion stars. Isolated stellar black holes are much more difficult to detect, and there may actually millions of them wandering around the Milky Way undetected.

The second category of black holes are the so-called supermassives like Sag A*. We now believe that almost all galaxies have at least one supermassive black hole in their center, with masses millions or even billions of times the mass of the Sun. These came as something of a surprise to the astronomical community—no one really expected to see anything like Sag A* in the sky. We really don't understand how supermassive black holes form or the role they play in the evolution of the universe.

Finally, we note that there is a big gap between the masses of stellar black holes and supermassives. Why there should be such a gap is another thing that we don't know about black holes.

## SEARCHING FOR EXTRAGALACTIC SUPERMASSIVE BLACK HOLES

Put yourself in the state of mind of an astronomer back in 1909 looking at some strange data coming from a distant nebula. In the first place, you would have no idea that there was such a thing as another galaxy out there, so you wouldn't interpret your data as emission from the heart of a distant galactic structure. In the second place, the modern concept of black holes was a decade into your future, so you certainly wouldn't be thinking about explaining your data in terms of supermassive black holes. Most likely you would just file your data under "To be investigated later," tuck it away in an unused corner of your office, and forget about it. Half a century later some hot shot kid will come along, look at that data, and wonder how you missed such an important piece of evidence.

This kind of "postdiscovery" process is fairly common in astronomy, as we saw with William Herschel's discovery of the planet Uranus. Once he announced his discovery, other astronomers looked through their back data and discovered that they, too, had sighted the planet but had thought it was a star or an asteroid. In fact, the postdiscovery data point played an important role in documenting the orbit of Uranus, a process that eventually led to the discovery of Neptune.

To understand what those old-time astronomers might have seen, we have to go into a little more detail on the process of spectroscopy, the process that allows us to establish the presence of atoms and molecules by looking at the light they emit. In a single atom, there are a set of discrete so-called allowed orbits in which electrons can move. Electrons can orbit the nucleus of the atom only in the allowed orbits. If an electron jumps from an outer orbit to an inner one, the difference in energy will be radiated away in the form of electromagnetic radiation of a specific wavelength. Similarly, if the electron absorbs electromagnetic adiation from the environment, it can jump from an inner to an outer orbit.

Combine all the different ways an electron can shift around among allowed orbits, and you can see that a given atom will radiate a collection of electromagnetic waves and that these waves will have a specific set of wavelengths. Since the allowed orbits are different for different chemical elements, you can think of this collection of wavelengths as a kind of "fingerprint" for each specific element. The collection of wavelengths emitted by a given element is called its "spectrum." A spectrum produced by electrons moving between outer to inner orbits is called an emission spectrum, while a spectrum with dark lines caused by the absorption of light as electrons move from lower to higher orbits is called an absorption spectrum.

In theory, when you take light from a given source, you can spread the different wavelengths out by passing that light through a prism. If you then look at the resulting spectrum, you should see a series of sharp lines at specific wavelengths, with each line corresponding to the movement of electrons between different allowed orbits. In a real situation, however, the atoms in the sample of material you are examining will be moving around—some moving toward you and some moving away from you. This means that, because of the Doppler effect (see chapter 7), the wavelengths of the light you see will be somewhat spread out. Some waves will have a bit shorter wavelength than the average and some a bit longer.

In very hot objects like the Sun, on the other hand, the atoms are moving around so much that the light winds up coming out at every wavelength, producing what is called a continuous spectrum—a spectrum that, like sunlight, contains all of the colors of the rainbow.

Sunlight, however, is not completely continuous, since there are gaps in its spectrum in wavelengths other than visible light. We know of no objects in nature that emit a continuous spectrum with no gaps from radio to gamma rays.

The scientist who may have seen a supermassive black hole without knowing it was the American astronomer Edward Fath (1880–1959). Working at the Lick Observatory near San Francisco, he undertook a study of what he called spiral nebulae but we would call spiral galaxies. Of the six nebulae he was able to observe, five looked to him like collections of ordinary stars, but in the sixth he actually saw an emission spectrum. Today, looking at his data, we realize that what he saw was the emission lines associated with the accretion disk of a supermassive black hole at the center of his galaxy. There was no way he could have imagined that his spiral nebulae were actually galaxies, that he was seeing material falling into a black hole, and that the black hole was millions of times more massive than the Sun. All of those concepts lay well in the future.

As the story of the discovery of SagA* demonstrates, the real gold standard that allows us to identify a supermassive black hole in the heart of a distant galaxy is data on the orbits of close in stars. This was the kind of evidence that really nailed down the identity of SagA*. If we are fortunate enough to be able to see these sorts of stars in another galaxy, we can use simple Newtonian mechanics to get the mass of the central object around which the stars are moving, From the mass we can calculate the Schwarzschild radius—the radius of the event horizon of a black hole of that mass.

Most of the time the resolution of the telescope being used is not fine enough to allow us to isolate a central black hole and its accretion disk. Light will be coming into the telescope from millions of stars, gas clouds, and other structures as well as the supermassive black hole. Astronomers have developed a number of complex mathematical processes to sort out the incoming radiation and estimate the mass of the central body that is producing the gravitational effects we are measuring. One of those techniques, Doppler broadening, helps illustrate how hard the problem of estimating the mass really is.

The simplest kind of sorting process depends on the fact that stars and gas clouds some distance away from the central mass will be moving at the sedate velocities associated with the rotation of the galaxy, while objects near that central mass will be moving much more quickly because of the intense gravitational field produced by that mass. We can then use the Doppler effect to analyze radiation from matter close to the central mass before it takes its final plunge into the black hole. If you think about matter moving close to the central mass, you can see two separate effects. First, as we pointed out in chapter 7, some of the material circling the central object will be moving away from you (and therefore its light will be redshifted), while some will be moving toward you (and therefore its light will be blue-shifted). In addition, the speed of matter spiraling around the central black hole increases as it gets closer to the central mass. Both of these processes have the effect broadening the profile of the spectral lines. The amount of broadening will depend on the speed at which the matter is moving and hence on the mass of the central object. The idea is that astronomers can observe a common spectral line—in light emitted by hydrogen, for example—and, from the amount of broadening, they can estimate the mass of the object around which the disk material is rotating.

Once we have an estimate of the mass of the central object our next job is to find its volume. Usually, as we pointed out, the resolution of our telescopes isn't up to the job of making this measurement directly, so we need to find another way to estimate the size of the central object. One measurement at our disposal is related to the fact that the light from some supermassive black holes at the heart of quasars is observed to change in brightness very quickly. It is thought, for example, that when a black hole swallows matter, or an entire star, there is a large "burp"—a sudden increase in brightness. These changes can happen very quickly, in days or even hours. If we know how quickly these changes occur, we can argue that the object doing the emitting can be no larger than distance light can travel in that time. It takes light approximately five hundred seconds (about eight minutes) to get from the Sun to Earth, for example, so if we see a change happening on that time scale, we can conclude that the object doing the emitting would fill the solar system out to Earth's orbit if it were placed in the position of the Sun. For reference, the Schwarzschild radius of a black hole whose mass is a billion times that of the Sun is about three light-hours across—enough to fit inside the orbit of Uranus if it were in the solar system.

The final step in the identification of an extragalactic supermassive black hole is to look at the amount of mass needed to produce the observed orbits at the galactic center, calculate the estimated volume, and ask if any object other than a supermassive black hole could fit those parameters. If the answer is no, we've identified another extragalactic supermassive black hole.

Over the years, these sorts of investigations have revealed several important facts about black holes:

- Galaxies like the Milky Way are littered with stellar-sized black holes, the end state of short-lived giant stars. Some estimates, for example, suggest that there may be millions or even billions of such black holes in the Milky Way.
- Almost all massive galaxies seem to have supermassive black holes in their center. We do not know how those black holes formed, a point to which we will return. Interestingly, the mass of the central black hole appears to be correlated to the mass of the entire galaxy itself, which has led to the widely held view that somehow the growth and evolution of the central black hole is connected to the

host galaxy in which it resides. These black holes must be connected somehow to the evolution of galaxies, but at the moment we do not understand that connection. We do know that the black hole exerts a profound influence on the host galaxy, ejecting material up to ten thousand miles (16,000 km) per second way out into intergalactic space. These massive energetic winds possibly suppress the formation of stars in the galaxy center resulting in a close symbiotic relationship between the feeding of the black hole and the regulation of star formation in the host galaxy.

- The bigger the galaxy, the bigger the central black hole. Again, we don't know why this should be so, but it seems to be the general rule.
- In general, black holes 100,000 times as massive as the Sun or more are considered to be "supermassive," while those 5 billion times as massive as the Sun are increasingly referred to as "ultramassive," although that term has not yet gained wide acceptance in the astronomical community.

## DESPAGHETTIFICATION

Despaghettification is probably not a real word, but we can't think of anything else to describe the subject of this section. In chapter 3 we described a piece of standard folklore associated with black holes—the process of spaghettification. This was, you will recall, the process that is imagined to take place when an observer approaches a black hole. There will be a difference in the gravitational force exerted on your head and feet, and that difference will eventually become big enough to tear your body apart, stretching it into long fibers (hence the name spaghettification).

There is no doubt that this process would occur if you approached a stellar mass black hole. Neil deGrasse Tyson described the process is gruesome detail in his book *Death by Black Hole*. You would expect, then, that supermassive black holes, with their huge mass, would be even more powerful producers of spaghettification.

They aren't, for roughly the same reason that Earth's gravity doesn't spaghettify you. The planet exerts a powerful gravitational force, but you are standing about four thousand miles (6,440 km) from its center. Remember that Newton's law of universal gravitation says that the farther you are from the center of a gravitating body, the weaker the gravitational force exerted by that body. Put in the numbers for a person standing near the event horizon of the billion solar mass black hole we described earlier, and you find that the difference between the gravitational forces of her head and feet is about the same as the difference between those forces when she's standing on the surface of the earth.

So much for spaghettification.

There are other unexpected properties of supermassive black holes. For example, if you divide the mass of some supermassives by the volume included inside their Schwarzschild radius, you find that the density of some black holes is less than that of water. Find an ocean big enough, in other words, and these black holes would actually float.

## THE BIG ONES

The supermassive black hole in Sag A* was hard for astronomers to accept: How could *anything* be that big? It wasn't long, however, before the search for extragalactic supermassives began to turn up much bigger candidates. As time went on, the question about Sag A* shifted from "How could it be so big?" to "Why is it only that big?" Consider the ultramassive black hole called TON 618, whose name comes from the fact that it is the 618th entry is a catalogue assembled at the Tonantzintla Observatory in Mexico.

One of the most massive black holes known, TON 618 weighs in at about 66 billion times the mass of the Sun. That means that you would have to take over half the stars in the entire Milky Way galaxy and scrunch them down into a black hole to match the mass of TON 618. It is located in a minor northern constellation near the Big

Dipper. Light from TON 618 has been traveling toward Earth for 10 billion years, an appreciable fraction of the lifetime of the universe. During that time the expansion of the universe has carried it farther away from the spot where the light was emitted; astronomers reckon that it is now about 18 billion light-years away from Earth.

TON 618 was first detected in 1957, well before people knew about quasars and supermassive black holes. It showed up in telescopes as a faint blue star. Hence, it is another example of what we called a "post-discovery" detection. Today we recognize that the light we receive from TON 618 is generated by matter falling into a huge black hole. It is likely that the central black hole is surrounded by a normal galaxy, made up of stars and gas clouds like the Milky Way. The radiation from the black hole is so intense, however, that any normal starlight that is being generated is completely blotted out.

The Schwarzschild radius of an object as massive as TON 618 is huge—the entire array of planets in our solar system would easily fit inside it. In fact, the Schwarzschild radius would put the outer edge of TON 618 forty times farther from the Sun than Neptune. All of which brings us back to a question we'll keep asking for the rest of this book: How did something this big come to be?

Astronomers are busy trying to peer into the center of galaxies with better and better spatial resolution. Adaptive optics (chapter 7) is now enabled on multiple telescopes throughout the world, allowing us to zoom into galaxy centers. The Very Large Telescope facility, as we have seen, combines four eight-meter (26.25 ft.) telescopes, which together can achieve the spatial resolution of an equivalent telescope of up to 130 meters (427 ft.) in diameter when looking in certain directions in the sky where bright enough guide stars are available. As we showed in chapter 7, gravity has delivered precision tests of Einstein's general theory of relativity and has provided strong experimental evidence that the compact mass in the galactic center (SgrA*) is indeed a Schwarzschild-Kerr black hole. In recent years, the most extreme resolution images of galaxy centers have finally been obtained with the Event Horizon Telescope.

## THE EVENT HORIZON TELESCOPE

By definition, we cannot see a black hole in the sense of seeing light that it emits—that's why they're called black. Nevertheless, we hope we have presented enough evidence to confirm their existence. Furthermore, when we talk about searching for supermassive black holes in other galaxies, we are talking about trying to detect objects millions of light-years away. No matter how big these objects are close up, at that distance they are far too small to be resolved by even the largest radio telescope.

Enter the Event Horizon Telescope (EHT), one of a number of cutting-edge systems capable of overcoming this problem and isolating central black holes in nearby galaxies. In essence, the EHT is a network of radio telescopes around the world—eight telescopes in the original buildout. This large collaboration, involving more than two hundred members and sixty institutes in twenty countries, achieves a resolution a thousand times greater than the Hubble Space Telescope. The easiest way to visualize the EHT is to think of the eight telescopes as constituting eight small pieces of one huge telescope the size of Earth. The idea is that the individual telescopes are equipped with atomic clocks that allow precise timing (one-tenth of a nanosecond) of the arrival of radio waves at each site. From this data the characteristics of the entire wave can be reconstructed.

As you can imagine, this sort of mathematical reconstruction is a pretty complex operation. For starters, the amount of information from each of the telescopes in the array is way more than can be sent over the internet. A typical night will yield about as much data as a year's worth of experiments at the Large Hadron Collider outside Geneva, Switzerland. It has to be sent in high-tech versions of flash drives carried on a 747 to centers at MIT in the United States and the Max Planck Institute in Germany, where the actual analysis is done. In computer slang, this is known as sending information via the "sneakernet," a name derived from the image of nerdy computer scientists in sneakers carrying data in their backpacks. If we used today's internet, to get data back from the South Pole, it would have taken twenty-five years.

The use of the sneakernet can cause some unusual problems. For example, one of the telescopes in the EHT is located at the South Pole. Since there are no flights to the South Pole during the Southern Hemisphere winter, the entire analysis program can be shut down while everyone waits for the South Pole data to arrive in the Southern Hemisphere spring.

The first images from the EHT were made public in 2019. They caused a minor media sensation, appearing on the front page of newspapers around the world. The picture wasn't of Sag A*, our own supermassive, but of the supermassive at the center of a neighboring giant elliptical galaxy known as M87.

As always, the astronomical nomenclature requires a little explanation. The French astronomer Charles Messier (1730–1817) published a catalogue of 110 objects that today we would call nebulae, galaxies, and star clusters. M87 was simply the eighty-seventh item in that list. It is a little more than fifty-three light-years from Earth—a near neighbor as these things go. Because of the problems with getting data from the South Pole, the analysis of M87 was finished before the analysis of Sag A*.

We can actually learn a lot about black holes by looking at the iconic picture of M87. We know that the path of light is bent by gravitational fields. (Recall that it was the bending of light rays by the Sun that provided one of the first experimental confirmations of general relativity in 1918.) The gravitational field of a supermassive black hole is so strong that some of the light generated in the accretion disc can actually be bent into a planetlike orbit. It is light escaping from this so-called photon ring that defines the inner edge of the visible dark spot.

That dark spot itself—the feature of the picture that attracted the most popular attention—represents the region where none of the photon trajectories escape the black hole. You can think of it as a kind of "shadow" of the supermassive black hole we are observing. The resolution required to see this shadow is remarkable. The shadow is only forty-two microarcseconds, equivalent to the resolution of holding out a single atom at arm's length or reading what's on a dime someone holds up in New York City if you're in Los Angeles.

The next thing you notice about the "photo" of M87 is the fact that there seem to be bright splotches of light in the ring around the central dark circle. This is the result of a process called Doppler beaming. It is something that happens when particles are moving near the speed of light as they emit electromagnetic radiation.

In chapter 7 we mentioned the fact that supermassive black holes often emit jets, typically from the poles of their axes of rotation. These jets have strong magnetic fields, and electrons moving at almost the speed of light corkscrew their way out around the field lines, emitting radiation as they go. This radiation (the technical term is "synchrotron radiation") can be thought of as something like a flashlight beam, with one beam moving in each direction away from the rotation axes. If we happen to be in the beam of the flashlight coming out from one pole, we will see a right spot. In this case, the darker areas will be produced by the jet moving in the opposite direction.

As of this writing, only two supermassive black holes have been "photographed" by the EHT: M87 and Sag A*. Presumably more super-massives will be added to this list as time goes on and more telescopes are added to the event horizon net. The most recent results from the

EHT were obtained using polarized light, which allowed the astronomers to map out the magnetic field generated by the black hole.

One final point: we have repeatedly made the claim that every galaxy, even small ones, has a supermassive black hole in its interior. Obviously this claim is not based on direct observation; we haven't examined every galaxy in the universe. The point, however, is that whenever we look at a galaxy, especially active galaxies, we find a situation consistent with the presence of a supermassive black hole. It makes sense, therefore, to take "every galaxy has a supermassive black hole" as a reasonable working hypothesis as we go forward.

# 9

# THE BLACK HOLE
# PARADIGM

---

**Merely corroborative details to lend an aspect
of verisimilitude to what would otherwise
be a bald and unconvincing narrative**

*W. S. Gilbert and Arthur Sullivan,* The Mikado

**M**ost galaxies are like the Milky Way. They are quiet, sedate, homey places where stars, for the most part, live out gentle lives as they burn through their allotted supply of fuel. There's the occasional supernova when a large star bites the dust, but a slow decline into white dwarf old age is much likelier. Even the supermassive black holes we know to be at the center of these calm galaxies are relatively quiet, or, to use the term preferred by astronomers, "quiescent."

A few of the galaxies we see out there are very different, however. Something wild is going on in the central nuclei of these galaxies. Huge amounts of energy are being thrown out, and many are shooting massive jets out from the galactic poles. In some cases the explosive events can outshine the entire galaxy from which it comes. All of this turmoil typically comes from the central nucleus of the galaxies. Astronomers refer to these systems as "active" galaxies and, since the energy is generated in the nuclei of these active galaxies, refer to the system as an "active galactic nucleus," or AGN for short. Our theories tell us that the huge amounts of energy we see are supplied by matter falling into supermassive black holes in the galactic nuclei. Roughly speaking, a few percent of the billions of galaxies in the universe are of this type.

Astronomers generally date the discovery of AGNs to a publication by the American astronomer Edward Fath in 1909 (see chapter 8). Progress in those days, since the data had to be obtained by detailed, tedious examination of photographic plates, was very slow. Nevertheless, in the spectra of systems that are now identified as AGNs, astronomers saw unusual spectral lines that were not characteristic of light emitted by stars.

There is a folktale among astronomers that when the American astronomer Vesto Slipher (1875–1969) was trying to measure light emitted from the nuclei of nearby galaxies, he would expose his photographic plate for an entire night, then put the plate into a dark box, stow the box in a dark cabinet, and take it out the next night. Sometimes he would have to repeat this process for several nights until enough photons fell on the plate to leave an image.

It wasn't until the advent of radio astronomy, however, that significant advances in the study of this kind of galaxy were made. Even so, it took a while for astronomers to connect AGNs to black holes. Perhaps the most significant work in this early period was done by the American astronomer Carl Seyfert (1911–60), who in 1943 published a catalogue of nearby galaxies that had the unusual kind of spectra that later came to be associated with AGNs. Galaxies of this type are now called Seyfert galaxies in his honor.

Our favorite step in understanding took place at a major scientific meeting in 1958. The Solvay Conferences are named after industrialist Ernest Solvay, who supplied the initial funding to get them started in 1911. They have always been elite, invitation-only events, and a lot of the science of the early twentieth century, like the famous Bohr-Einstein debates on the nature of quantum mechanics, took place at these conferences, which are still being held in Brussels.

In 1958 the Armenian astronomer Viktor Ambartsumian (1908–96) gave a presentation on AGNs that included this statement: "Explosions in galactic nuclei cause large amounts of mass to be expelled. For these explosions to occur, galactic nuclei must contain bodies of huge mass and unknown nature." You couldn't ask for a better description of a supermassive black hole.

Over the last half of the twentieth century we began to accumulate knowledge about all the AGNs we see out there. As more data came in, patterns began to emerge that allowed us to set up a classification scheme for these explosive objects. And this brings us to a question that scientists always run into when they start the classification process. The problem usually comes up in biology, but it definitely came up with AGNs as well.

Here's how it works. Suppose you find two AGNs that look similar but differ in some respect—one might emit twice as many radio waves as the other, for example. There are two ways that we can deal with this situation, often referred to as "splitting" and "lumping" A lumper looks at this situation and decides that there is only one category with some variation. A splitter, on the other hand, would insist that these objects represent two separate categories and should be regarded as completely separate.

The literature on AGNs is full of splitters and lumpers. In our background reading for this chapter, for example, we found classification schemes for AGNs that had fewer than a dozen entries and one that had no fewer than twenty-five. We tend to be lumpers, but it's important to realize that these differences have nothing to do with the physics of AGNs and are more involved with semantics than with the physical reality.

## PROPERTIES AND TYPES OF AGNS

We can start with some general properties we see in many AGNs. For example, we can ask about the source of the enormous energy all AGNs emit. If we generate energy from chemical reactions—for example, burning anything from wood to rocket fuel—we recover about 0.01 percent of the energy, including the mass energy, in the fuel. This is about the maximum you can get by shifting electrons around. A device that generates energy from nuclear fusion reactions, the source of energy in stars, will get you roughly 0.2 percent of the energy in the mass of your fuel. Allow that fuel to fall into the accretion disk of a black hole, and you will get anywhere from 5 percent to 40 percent of the mass energy. This is the most likely source of the intense energy coming out of AGNs. So the first conclusion we can draw is that AGNs are likely powered by matter falling into supermassive black holes.

Now let's look at the radiation emitted by AGNs. We can start by noting that the radiation from any AGN can be quite complicated. Different regions of the nuclei send out different types of spectral lines. Some lines are clearly the result of Doppler broadening (see chapter 8), indicating turbulence and high temperatures, as well as the fact that the material emitting these lines is rotating in an accretion disk, which means that sometimes it is moving toward us and sometimes it is moving away. To be precise, these broad lines do not originate in the accretion disk itself but from material outside the disk. Other spectral lines are narrow, indicating that they originate in a relatively calm, low temperature region. As we stressed earlier, these sorts of spectral lines cannot originate in stars, which strengthens our conclusion that stars are not the source of the AGN's energy.

And then there are jets. Jets are huge structures that dwarf the galactic nucleus (and, for that matter, the entire galaxy) where they originate. They typically come out back-to-back in a direction perpendicular to the accretion disk. Jets are streams of relativistic particles that can extend for hundreds of thousands of light-years away from their source—think of them as clouds of electrons corkscrewing their way along the jet, emitting radio waves as they go. We don't really understand how jets form, but they seem to be found in a number of AGNs.

There is one more feature of normal black hole structure that makes sense but nonetheless seems strange. We know that gravitational forces can deflect the path of light. Indeed, the observational verification of this fact was the reason that general relativity was so easily accepted by physicists. It turns out that as we get close to the event horizon of a black hole, the gravitational force can become so strong that incoming photons are bent into circular paths. In effect, they go into orbit in much the way a planet goes into orbit around a star. These so-called photon orbits produce a luminous path around the black hole.

Having made these general comments about black hole structure, we can go on and look at specific examples of AGNs.

- The *quasar* is probably the most interesting AGN from a scientific point of view.
- The *blazar* is in the same family as quasars. A blazar is simply an AGN one of whose jets is pointed directly at Earth, so that we are, in effect, looking down the barrel of a gun. In a situation like this, a very small change in the direction or density of the jet can produce

a big change in the intensity of the radiation we see. Like all AGNs, blazars are powered by a supermassive black hole. They appear very bright in part because the electrons in the jet are moving at relativistic speeds, and hence undergo a brightening process known as relativistic beaming.

- *BL Lac objects* are members of the "it's like a quasar" family. We include them here just because they have an interesting history. As far back as 1928 a BL Lac object was included in a catalogue of variable stars. At the time the object was thought to be in the Milky Way, and it was given the name of BL Lac for the constellation in which it appeared, Lacerta, the Lizard. Subsequent observations, however, proved puzzling. For example, variable stars usually go through a regular brightening and dimming sequence. BL Lac did get brighter and dimmer, but in no discernible pattern. Furthermore, it seemed to be radiating a lot in the radio region of the spectrum, which was unusual for a star. Eventually it was realized that it wasn't a star at all but the nucleus of a distant galaxy. The radiation from these objects is characterized by a flat spectrum with few spectral lines.

- *Seyfert galaxies* have the characteristic intense radiation of AGNs but generally emit less radiation than quasars. Also, unlike the case with quasars, it is often possible to see the galaxy in which a Seyfert nucleus resides. Seyfert galaxies are quite common, perhaps amounting to 10 percent of known galaxies. Astronomers generally characterize Seyfert galaxies according to the kind of spectral lines seen in their spectra.

In chapter 8 we discussed the phenomenon of Doppler broadening, in which broad spectral lines are emitted by regions in which atoms are going through rapid motion, often in many directions. Narrow lines, on the other hand, originate in regions where emitting atoms are relatively quiet. Seyfert galaxies come in two "flavors": class I, in which both broad and narrow lines are seen, and class II, in which we see only narrow lines. (We note that splitters have offered more complicated categorization schemes for Seyfert galaxies.)

## UNIFICATION MODELS

Obviously, the categorization of AGN has led us into a veritable jungle of complexity. It has been suggested that this complexity may not represent real physical difference between different types of AGN but is instead caused by the fact that we are observing complex systems from different points of view. The complexity, in other words, must arise because we are looking at a single system in different ways.

Let's start by putting together all the pieces of an active galaxy system. At its core is a supermassive black hole. The black hole is surrounded by a hot accretion disk that feeds matter into the black hole and generates the energy we observe. This entire system is encased in a dusty torus, cool and full of gas and dust. There may be powerful jets shooting far into space, presumably along the direction of the rotation axis of the black hole. This paradigm is still much debated, and refinements to this simple model continue to be made.

Imagine now that the accretion disk is oriented perpendicular to the sheet of paper you are reading right now and that one of the jets is moving toward the top of the page. This is a complex system with many parts. It will look different to different observers. We can start with an observer at the top of the page, right in the path of the jet, who will see a blazar.

Now move away from the jet in a counterclockwise direction. Assume that your line of sight to the black hole passes above the dusty torus. In this case you will see both broad lines (from gas near the accretion disk) and narrow ones (from colder material above the torus). To you this object will appear to be a class I Seyfert galaxy. Keep rotating in a counterclockwise direction. Eventually your line of sight to the accretion disk will be blocked by the dusty torus. Now you will see only narrow spectral lines (from the torus), and you will classify the object as a class II Seyfert galaxy.

# 10

## QUASARS

————

**If the radiance of a thousand suns were
to burst at once in the sky, that would be
the splendour of the mighty One.**

Bhagavad Gita *11:12*

I n the mid-twentieth century, astronomers kept running into situations that could be explained only by the presence of black holes.
The long history of the discovery and explanation of quasars is a
case in point.

"Quasar" is a shortening of the phrase "quasi-stellar radio
source." It refers to objects that are emitting large quantities of radio
waves but are relatively small, often about the size of a typical star.

In chapter 6 we talked about the beginnings of radio astronomy
in the 1930s, which opened a new window on the universe. The first
thing that scientists usually do when a new territory opens up like this
is do a survey, a scientific Lewis and Clark expedition of a kind under-
taken to give us a rough map of newly discovered terrain. The first

radio surveys had another important use, for the turn of the twentieth century saw a massive increase in the use of radio communication. If you're sending out a radio signal, it is very useful to know if there are natural sources of radio waves that might interfere with that signal. In the 1960s, similar concerns about microwave communication led to the discovery of the cosmic microwave background, one of the most important pieces of evidence supporting the Big Bang theory.

The radio surveys of the sky began turning up some strange radio sources. These sources were very bright in the radio sky, but astronomers had a hard time associating them with any sources of visible light. Their luminosity sometimes changed quickly, indicating that they were small—perhaps no larger than our own solar system. Finally, when optical identification of quasars was finally established, the spectra didn't look like anything associated with atoms we know about on Earth.

By 1959 hundreds of these strange beasts had been identified, and astronomers entered them in a publication known as the Third Cambridge Catalogue, in which a specific quasar was at first referred to by a title such as "3C273," meaning the 273rd entry. Once it became clear that there are a huge number of quasars in the sky, this system was replaced by one in which quasars are referred to by their position in the sky, rather than their place in the catalog, which contained no fewer than 470 radio sources.

The publication of the catalog was a crucial point in the history of astronomy. For the first time astronomers were doing a systematic survey of the heavens using something other than visible light. As we have noted, Earth's atmosphere blocks all incoming electromagnetic radiation except for visible light and radio. The universe may have been sending us all other kinds of signals, but they weren't getting through. The Cambridge Catalog, then, represents the first look at a different universe, one sketched by radio waves. This was followed in the second half of the twentieth century by astronomical surveys of the sky in all the rest of the electromagnetic spectrum: microwaves, infrared and ultraviolet light, X-rays, and gamma rays. Each of these new explorations required a new technology and new types of instruments, as different from optical telescopes as are radio dishes. In most cases these new explorations could be done only with receivers above Earth's atmosphere. Consequently, there are dozens of astronomical satellites in orbit today, catching the incoming radiation before it is destroyed by our atmosphere.

The discovery of quasars triggered a major debate in the astronomical community. The central issue involved the fact that they were very bright in the radio sky. There are only two ways to achieve this kind of brightness. One is for the source to be close, probably within our own galaxy. The other is for the source to be far away but tapping an as yet unexplained energy source that allowed it to put out such a prodigious amount of radiation.

It wasn't long before astronomers began detecting visible light coming from already discovered quasars. Of particular interest is quasar 3C273. Today we know that it is more than 2 billion light-years away, well outside the Milky Way. This wasn't known in 1962, however, a year in which the Moon was scheduled to pass between Earth and 3C273 several times. Because the position of the Moon is known very accurately, it was possible to make detailed measurements of the quasar as it slowly came back into view from behind the Moon. The Dutch American astronomer Maarten Schmidt (1929–2022), working with the telescope at Mount Palomar, California, was able to obtain spectra of the material around the quasar. He realized that the "anomalous" lines were just emissions from ordinary hydrogen but shifted far into the red because the quasar was receding from us at the astonishing speed of 29,200 miles per second (47,000 km/sec), one sixth of the speed of light. No stars were known to move that fast.

Another set of measurements was made by astronomers in Australia and New Zealand. They positioned a radio telescope on a high cliff overlooking the ocean between the two countries and made detailed measurements of quasars as they rose above the horizon and sank into it. This allowed them to estimate the celestial coordinates of each quasar in their sample, an important step in determining its true position and distance from Earth.

So now what to do with these measurements? If a quasar was close—say, inside the Milky Way—then it would have to be made of stuff that didn't exist on Earth. How else to explain the strange spectral lines? If a quasar was really at cosmological distances, as it actually turned out to be, how could it be so bright? It would have to be pumping out more energy than all the stars in the Milky Way taken together. In 1962 no one had any idea that that amount of energy could be produced by celestial objects.

One of us, Jim Trefil, had an interesting experience during this period of turmoil and debate. He was a speaker at a scientific conference in England, a conference at which Maarten Schmidt was the keynote lecturer. After the formal proceedings were over, Schmidt sat down to have a beer with a group of us young scientists, one of those informal experiences that are the main benefit of scientific conferences. What impressed Jim about Schmidt's recounting of his work was not so much technical difficulties as the fact he sat on his results for *months* while he checked through his measurements and, in Richard Feynman's words, made sure that he wasn't fooling himself. Some modern scientists today would be on the phone to a *New York Times* reporter before the computer had finished spitting out the data. Schmidt, one of the few scientists to appear on the cover of *Time* magazine, was definitely old school.

What Schmidt showed was that quasars are made of ordinary stuff and very far away. Where could all that energy come from? In chapter 3 we introduced the idea of black holes having accretion disks. Think for a moment about how much energy matter falling into a black hole is capable of generating. Imagine, for example, that Earth suddenly changed into a marble-sized black hole. You would be four thousand miles (6,437 km) from the event horizon, and you would start falling. By the time you got to the event horizon you would be traveling at an appreciable fraction of the speed of light. That's a whole lot of energy. In a real situation you probably wouldn't fall straight in but would spend some time in the accretion disk, which doesn't change the fact that an object falling into a black hole is going to generate a lot of energy before it disappears. If you imagine a supermassive black hole swallowing whole star systems along with all the interstellar gas in its vicinity, it's not hard to see how that black hole could outshine a typical galaxy like the Milky Way.

The usual analogy that is used to describe this behavior is a rather messy eater consuming whatever matter might come its way and converting it into radiation. This analogy makes it easy to describe the observed rapid brightening and dimming that we sometimes see in quasars. They appear as the result of a sudden increase in matter falling past the event horizon. A whole star, for example, might brighten the radiation temporarily. It is customary, by the way, following our gustatory analogy, to call this a "burp."

By 1969 the idea that supermassive black holes at the heart of nascent galaxies are generating the enormous energy output of quasars was taking hold. Quasars, in other words, are just the first stage in the formation of galaxies. And since quasars are powered by matter falling into supermassive black holes, this means that supermassive black holes are somehow connected to the formation of the familiar universe. We still have to work out the details of exactly how this works. It suggests, however, that our familiar universe, and life itself, may depend on the behavior of supermassive black holes. One interesting speculation follows from this idea: if quasars are really an early stage in galactic evolution, then someone looking at the Milky Way from a distance of billions of light-years might see a quasar where our galaxy is today.

The "burps" in the quasar output—rapid changes in luminosity, to be more formal—tell us that on a cosmic scale, quasars are quite small. Most estimates, in fact, come up with sizes for these objects less than the size of our own solar system. And while this might seem quite large to us, compared to interstellar distances it is miniscule. Again we have a mystery. Whatever is producing the radiation has to be small, yet it has to be incredibly powerful. Again, supermassive black holes are just about the only thing we can think of that could satisfy both of these requirements. Calculations suggest that up to 40 percent of the mass of material falling into a black hole can be converted into energy, a much more efficient process than the nuclear fusion that powers most stars.

As of this writing we have seen more than 4 million quasars and counting. Thanks to many new sensitive large sky surveys, the number of new quasars discovered has increased exponentially, with hundreds being discovered at very high redshifts, within the first billion years of the universe's history. The ubiquity of these objects is strong evidence that they must play an important role in galactic evolution.

# 11

## GRAVITATIONAL WAVES

---

### There are no gravitational waves analogous to light waves.

Albert Einstein, 1916 letter to Karl Schwarzschild

**A**lbert Einstein couldn't be right all the time. He had a fraught relationship with the concept of gravitational waves, often acknowledging their existence and denying it in publications less than a year apart. Since the ability to detect gravitational waves is evolving into one of the best ways to get information about black holes, it's probably a good idea to get a sense of how they can be seen as simple consequences of general relativity.

In chapter 2 we introduced general relativity using the example of a bowling ball dropped onto a rubber sheet. Those of us who teach introductory relativity love this "bowling ball on the trampoline" analogy, simply because students always seem to get it when the analogy is used. As it happens, the bowling ball and the trampoline also give us a simple way to see how general relativity leads to the prediction of gravitational waves.

Imagine putting the bowling ball down on the trampoline and then grabbing it and pushing it rapidly up and down. The motion of the bowling ball will produce an outgoing ripple in the fabric of the trampoline. This ripple is the analogue of a gravitational wave. According to the theory, whenever a mass is accelerated, one of these ripples in space-time goes out. Because gravity is the weakest of the fundamental forces, these ripples are small and extremely difficult to detect, but they are definitely there.

Although Einstein is generally credited with predicting the existence of gravitational waves despite his obvious skepticism, we have to note that ideas that eventually led to gravitational waves were in the air in the late nineteenth century. One reason, as we saw in chapter 1, was the old problem of the "spooky action at a distance" connected with Newtonian gravitation—it even bothered Newton that two objects that were not in contact could exert forces on each other. He famously replied to critics worried about how gravity was transmitted from Earth to the Moon with the Latin phrase *Hypotheses non fingo*, literally, "I frame no hypotheses," which translates to something along the lines of, "Frankly, guys, I haven't the faintest idea of how the force is transmitted, but the equations seem to work."

The idea that some sort of medium carried the force between massive objects was popular at the time because it seemed to echo the generation of electromagnetic waves connected to Maxwell's equations. If accelerating electrical charges can produce electromagnetic waves, the argument went, why shouldn't accelerating masses produce gravitational waves? There is, however, an important difference between gravitational and electromagnetic waves, even though both travel at the speed of light. Electromagnetic waves are electrical and magnetic field traveling *through* space. Gravitational waves are traveling distortions *of* space. Space-time, in other words, is the stage upon which electromagnetic waves move, while is it the motion of space-time itself that constitutes a gravitational wave.

Imagine a stadium full of people. One way to move information around the stadium would be to have people pass a note, each person handing it on to their neighbor. This would be a wave analogous to the electromagnetic wave traveling through space-time but not affecting it. Now imagine people doing the wave, where each person stands up as the wave reaches his or her position. This would be the analogue of a gravitational wave being a distortion of space-time itself. In the first example space-time is a sort of neutral stage through which a message passes. In the second it is the changes in space-time itself that produce the wave.

Having made this point, we can go on to ask what a passing gravitational wave would look like. Suppose we start with a spherical object, like a basketball or a soccer ball. As the wave went by, we would see the shape of the ball change. First it would be distorted into the shape of a football, with its long axis in the horizontal. Then it would return to the original spherical shape. Then it would be distorted into a football with is long axis vertical, and, finally, return to the original spherical shape. Seeing this distinctive set of changes would be a positive signal that a gravitational wave had passed by our basketball.

How big will those distortions be? How accurately would we have to measure the dimension of that basketball to be certain we had seen it go through the contortions we outlined? Here's where we get into trouble. The basic problem is that because gravity is by far the weakest of the fundamental forces, it takes a whole lot of mass to produce even the smallest gravitational wave. In principle, every time you wave your hand you are creating a gravitational wave, but that's not a wave anybody could imagine detecting. Even our most complex measuring instruments can see only waves that are created by the collision and merging of two stellar mass black holes. And even these waves require us to detect changes in the basketball's diameter that amount to a tiny fraction of the distance across an atomic nucleus. This is a feat equivalent to measuring the distance from New York to San Francisco to within a thousandth of the width of a human hair.

The person generally recognized as the founder of the field of gravitational wave detection is University of Maryland physicist Joseph Weber (1919–2000). Before he settled down as a university professor, Weber led an unusually interesting life. He attended the U.S. Naval Academy and was involved in many of the naval battles of World War II. He delighted in regaling his students with an account of his adventures during the Battle of the Coral Sea, where his ship, the USS *Lexington*, was sunk after destroying a Japanese aircraft carrier. After the war Weber remained in the navy, retiring with the rank of commander, before joining the faculty of the University of Maryland. Academics are often surprised to learn that he received his faculty appointment before he finished his PhD.

In 1955 Weber received a Guggenheim fellowship that allowed him to spend time at the Princeton Institute for Advanced Study. It was this opportunity to think about general relativity that led to his lifelong attempt to detect gravitational waves. Weber's method of detection started with the notion, which we discussed, that passing gravitational waves will distort solid objects that they encounter. A metal object will deform as the wave goes by, then snap back into its original shape when the wave passes. This process will set up sound waves that reverberate in the metal object, sound waves that can be picked up by sensors attached to the outside of the object. Weber's scheme was to set up a pair of aluminum cylinders about 6.5 feet (2 m) in length and a meter in diameter, one in College Park, Maryland, and the other at Argonne National Laboratory outside of Chicago. He calculated that a passing gravitational wave would change the length of each bar by about $10^{-16}$ meters, about the same change in length that one would expect from normal thermal oscillations in the bar. Weber argued that seeing the same signal in the two detectors would be evidence for the detection of a gravitational wave, and in 1969 he announced that he had been successful. He traveled around the country, explaining his finding to university and laboratory audiences. Jim Trefil remembers him showing a slide that had a huge spike in the aluminum bar's oscillation, a spike that Weber jovially explained was not a gravitational wave but an incident in which a National Science Foundation executive had backed his car into the Maryland physics building. Needless to say, this spike didn't show up at Argonne Labs.

Unfortunately, other groups tried without success to replicate Weber's results. The reason seems to have been his inattention to a rather obscure point of statistics. Here's a simple explanation: suppose you and a friend agree to do an experiment in which you each roll a set of dice a hundred times, always initiating the throws at the same time. If you look at the result, you will find that your and your friend's dice will give the same result for some throws—you might both throw 6 on throws 14, 56, and 78, for example. This does not mean that there is a physical cause for those instances of 6 to show up. It's simply a matter of probability. In a Weber-type experiment, the sensors attached to the aluminum bars will be constantly giving varying signals due to the normal thermal motion of the bar's atoms. Occasionally, some of these signals will be a bit higher than normal. Think of those elevated signals as being analogous to the throws of the dice in our example. By chance, it will occasionally happen that one of these enhanced signals will appear in both detectors, just as occasionally you and your friend will get the same number on your dice. Working out how many times this sort of statistical effect can mimic the kind of result you're looking for in an experiment is one of the hardest tasks that experimental scientists face. The general consensus in the physics community is that Weber just didn't get this aspect of his results right, so that basically what he detected was a statistical fluctuation and not a gravitational wave.

Joseph Weber may not have detected gravitational waves, but he is honored today as the first scientist to attempt this difficult task, putting gravitational wave detection on the map. There is a Weber Memorial Garden on the campus of the University of Maryland that features, along with the flowers, some of those aluminum bars that got the whole field going. The one lesson we can take from the Weber affair is that any detection of gravitational waves is going to require the ability to measure very small displacements of matter. Enter the interferometer, a nineteenth-century apparatus that, combined with modern lasers, does exactly that.

An interferometer works by harnessing the nature of light waves. When two light waves arrive at the same spot, they exhibit a type of behavior called interference. All waves—sound waves, earthquakes, waves on the ocean—do the same. Suppose you have two have sources of water waves, such as might result from having a pair of paddles periodically hitting different places on the water surface. Suppose further that both wave sources are ten feet (~3 m) away from you—one on the left, the other on the right. Finally suppose that the waves from each of the sources arrive at your position so that the crests of the two waves show up at the same time and reinforce each other. What you will see, then, is a large wave. We say that you are seeing "constructive" interference between the waves. If, however, the situation was such that the crest of one wave arrived at your position at the same time as the trough of the other, the two would cancel each other out. This is known as "destructive" interference. Obviously, between these two extremes there are situations where there is partial destruction and partial construction.

If we move the source of one of the waves, gradually increasing the distance between the two sources to one wavelength of the wave that is being generated, the interference pattern will go through its entire cycle from constructive to destructive. This means that by examining the interference pattern created by your two waves you can tell how much the moving wave source has been displaced.

Let's replace the water in our example with light waves produced by an ultraprecise laser, a single laser source whose output is run through a device called a beam splitter. The function of the beam splitter, as the name implies, is to produce two beams of light from the original beam, typically by running the original beam through a special crystal. At this point we have the basic working parts of a modern interferometer. In a typical experiment we might send the two beams off in mutually perpendicular directions and let them bounce off of mirrors and come back to the observation point. By measuring the interference pattern formed when the two beams recombine, we can detect small changes in the distances between the beam splitter and the mirrors.

One of the most famous experiments involving an interferometer was done in 1887 at what is now Case Western Reserve University in Cleveland. Physicists Albert Michelson (1852–1931) and Edward Morley (1838–1923) set out to measure the effects of a substance called ether that people believed filled the space between physical objects. They reasoned that the movement of the earth through the ether would create an "ether wind" that would affect the movement of light in the arms of their interferometer. (The argument is similar to having a boat moving with and against the current in one arm and across the current in the other.) When they failed to detect any signal, they forced scientists to conclude that the ether simply didn't exist, a result that was later taken to support the theory of relativity.

A much more ambitious interferometer began to take shape in the minds of some physicists in the 1960s. Einstein's initial skepticism about gravitational waves had been based largely on the fact that the only objects capable of producing these waves known to science in 1916 were far too small to produce anything that could conceivably be measured. Why bother to speculate about something you could never see? We mentioned that, at least in principle, you can create a gravitational wave simply by waving your hand. This statement is true, but the wave you create in this way is so weak that it couldn't be detected by any conceivable measuring system.

By the 1960s and '70s, however, we had discovered objects in the universe that were so massive that their movement could easily create measurable gravitational waves. Neutron stars are one example, and black holes are another. These discoveries, totally unexpected in Einstein's time, had the effect of lowering the sensitivity needed to detect the (now much stronger) gravitational waves.

In the 1960s Weber's experiment was still on people's minds, and some groups tried to improve them by, for example, cooling the aluminum bars. Others, however, tried a completely new technology and hunt for the waves using an interferometer. Groups in Germany and at MIT built prototype detectors several meters in size. During this period two separate groups grew up in America. One, headed by physicist Rainer Weiss at MIT, was involved with building prototype interferometers and analyzing the sources of error in the instruments. The other, at Cal Tech, was headed by theoretical physicist Kip Thorne and examined sources of gravitational waves. The two institutions collaborated on the proposal to build a detector in 1987, and Weiss, Thorne, and Caltech physicist Barry Barish, who basically made the system work, eventually shared the 2017 Nobel Prize when their work eventually resulted in the detection of gravitational waves.

The real turning point came in 1986, when National Science Foundation convened a weeklong meeting in Washington, DC, to take a serious look at what kind of interferometer would be needed to detect waves from astronomical sources. The committee report was unambiguous: if we're going to be serious about detecting gravitational waves, let's go whole hog and build a pair of L-shaped interferometers with arms nearly 2.5 miles (4 km) long. This report can be considered the birth of LIGO (Laser Interferometer Gravitational Wave Observatory).

Why do interferometers have to be so big? The answer to this question comes down to the weakness of the gravitational force. Suppose you were measuring an effect that changed the length of a solid rod by 1 percent, so that a rod one meter (100 cm/3.38 ft.) long would shrink by 1 centimeter. A rod one hundred meters long, on the other hand, would shrink by one meter. If the only measuring instrument available to you could measure to the accuracy of only one meter, you would have to build an apparatus that was a hundred meters long to see the effect you were looking for. The size of your apparatus, in other words, depends on two things—the size of the thing you're trying to measure and the accuracy with which you can make a measurement. It was the combination of these two things that determined LIGO's size.

In reality, the LIGO system has to be able to detect movement in the mirrors amounting to one thousandth of the width of a single proton. Considering that the mirrors weigh about a hundred pounds (45.35 kg) each, this is quite a feat. Physicists who are working on increasing LIGO's sensitivity say that they are starting to run into fundamental limits imposed by the operation of quantum mechanical effects like those imposed by the uncertainty principle, which states that we cannot know the position and velocity of an object with perfect accuracy at the same time.

In the end, two sites were chosen for interferometers, one in Louisiana and the other in Washington State, about 1,120 miles (1,800 km) apart. Only signals that registered on both interferometers would be accepted as candidates for the detection of a gravitational wave. No officials backing their cars into a building would get through the LIGO screening process.

The builders of the LIGO systems were moving into uncharted territory, a fact that was reflected in the master plan. The first version of the machine would use well accepted and highly dependable technologies in a kind of dry run to make sure that the system was working. The machine would then be shut down and upgraded with new technology that was expected to be developed. The original LIGO came online in 1999 and ran unsuccessful searches for gravitational waves from 2002 to 2010. Then the "advanced" LIGO (aLIGO) came online in 2015 and in a matter of days registered its first wave detection. September 14, the date of the first detection, is thus generally taken to be the date when gravitational wave astronomy was born. More than 1,200 scientists from all around the world worked on various aspects of LIGO, representing more than a hundred different institutions, a wonderful example of international collaboration.

The first gravitational wave event was caused by a collision between two black holes estimated at 29 and 36 solar masses. Those masses pose a puzzle. The problem is that most stellar black holes we know about weigh in at just a few solar masses, with very few being more massive than 10 solar masses. Since this event was, as far as we can tell, completely random in nature, it's natural to ask why it involved such atypical masses.

We can think of several ways that this question might be answered. What we witnessed might be a step in the formation of supermassive black holes. Both participants, for example, might themselves be the product of previous collisions. On the other hand, it might be that when these particular black holes formed billions of years ago, black holes just happened to be bigger than they are today. We could probably assemble a long list of speculative answers to the size problem, but since none of them would be testable it's probably just best to file them in the "To be solved" folder and move on.

We also have to stress the tremendous amount of energy that was released in creating that wave. Astronomers estimate that this single collision released fifty times the amount of energy being released as electromagnetic radiation emitted by all the stars in the observable universe.

An important insight into black hole collisions as well as to the problem of intermediate mass black holes mentioned earlier occurred on May 21, 2019, when our gravitational wave detectors received the most powerful wave ever seen. It was apparently generated by the spiraling of two black holes of 85 and 66 solar masses, respectively, a process that produced a single black hole with a mass 142 times that of the Sun. This is in the middle of the intermediate mass black hole range. Theoretical calculations suggest that a mass equivalent to eight times that of the Sun was converted directly into the energy of outgoing gravitational waves, making this one of the most powerful events observed.

Even more than the advent of radio astronomy discussed in chapter 10, this new way of looking at the universe was truly revolutionary. Up until September 14, 2015, every bit of information we had about the universe with the exception of cosmic rays came to us in the form of electromagnetic radiation. In essence we saw what the universe was offering us in forms created by the movement of electrical charges. Now, for the first time, we were seeing information created by the movement of mass instead of charge. A whole new universe had suddenly opened up to us, a universe of gravitational waves that could be traced all the way back to the Big Bang itself.

An interesting aside is that LIGO scientists often presented their event to the public as an audible "chirp." This was possible because the frequency of the waves detected by LIGO, if converted into sound, lie within the human range of hearing. The chirp starts at 35 Hz, near the lower range of human hearing, and moves up to about 250 Hz as the black holes collide. (250 Hz is a bit lower than the pitch of middle C on a piano.) The little variations in the chirp actually carry information about the black holes whose collision produced the gravitational wave in the first place.

LIGO may have been the first operational gravity wave detector, but it was certainly not the last. Today LIGO's output is supplemented by gravitational wave detectors located in Italy and Japan, and this cadre will soon be joined by a detector in India. The presence of multiple detectors allows astronomers to do some triangulation to locate the source of the waves, information used to direct the attention of other telescopes to the newly discovered hot spot.

Before moving on to the immediate future of gravitational wave astronomy, there is one more point we need to make. Up to now we have discussed waves generated by the collisions of black holes, but these are not the only possible sources of gravitational waves. Any compact massive colliding bodies will produce the waves as well. In fact, LIGO has detected waves generated by collisions between black holes and neutron stars and collisions between two neutron stars as well as those between pairs of black holes. (The two neutron star events are increasingly referred to as "kilonovae.")

The first detection of a collision between two neutron stars generated a certain amount of attention in the popular press. To understand why this was so, think back to our discussion of the generation of chemical elements in stars, with a simple progression up to iron. The tradition was for astronomers to argue that all the heavier elements were forged in the maelstrom created by large stars going supernova. The problem was always that this sort of explanation never quite fit the data. The energy available in the collision of two neutron stars and their collapse into a black hole, on the other hand, was more than ample to produce elements all the way to the end of the periodic table. The *Wall Street Journal* latched onto the story of this collision because collisions like it are probably a major source of gold on Earth, a topic of obvious interest to their readers.

One problem that will have to be addressed as gravitational wave astronomy goes forward concerns the expected size of gravitational waves coming from sources other than the collision of relatively large black holes. Theorists have calculated the gravitational waves that would be produced by the most massive movement of matter imaginable, the Big Bang itself. Presumably these would be produced by all sorts of processes—supernovae, black hole collisions, and so on. These events would have produced a random collection of gravitational waves that have been sloshing around the universe (and expanding with it) since the earliest times. They would form a kind of gravitational wave background carrying information about the formation of the universe, much as the cosmic microwave background (composed of electromagnetic waves) contains information about the state of the universe when atoms formed. Astronomers call this a "stochastic" background because the waves have been generated by random events. (The word "stochastic" comes from the Greek word for "guess.") These waves, having been traveling and spreading out for so long, are well below the detection capability of LIGO.

As noted, one way to increase the sensitivity of an interferometer is to increase the length of its arms. We can see this in the difference between the tabletop experiment of Michelson and Morley and the four-kilometer arms of LIGO. There is, however, a limit to how far you can increase the arms of a detector on Earth's surface. The current dimensions of LIGO probably represent a practical and financial limit to the length of an interferometer's arms, but it's hard to see how they could be more than a few thousand miles long in any case. This raises a problem, because an interferometer of this unbelievable size would not be sensitive enough to see the stochastic gravitational waves generated just after the Big Bang.

If you can't build the detector on Earth, then you have to build it in space. Enter evolved Laser Interferometer Space Antenna (called eLISA, or usually, just LISA), a project being run by the European Space Agency (ESA). LISA will be composed of three satellites that define the vertices of an equilateral triangle whose sides measure 620,000 miles (1 million km) each, meaning that the combined sides of the triangle will be several times greater than the distance between Earth and the Moon. Each spacecraft contains two telescopes, two lasers, and two test masses of roughly 4.4 pounds (2 kg). The telescopes and lasers in each spacecraft are arranged in optical assemblies pointed at the other two spacecraft. The net effect is that they form interferometers like those used in the Michelson and Morley experiment, each centered on one of the spacecraft, with the test masses defining the ends of the interferometer arms. The idea is that a passing gravitational wave will change the lengths for the sides of the triangles, which the satellites will detect. This entire assembly will be put into orbit around the Sun, trailing along with Earth.

Obviously, an ambitious project like this depends critically on our ability to control the position of distant spacecraft. In 2015 the ESA launched the LISA Pathfinder, a proof-of-concept mission to show that it was possible to maintain this sort of control on a spacecraft. In the Pathfinder spacecraft were two test masses, located fifteen inches (38 cm) apart, whose position was monitored by a laser interferometer. ESA scientists quickly showed that the two test masses could maintain their relative position to an accuracy five time better than what would be needed in LISA. Its mission completed, LISA Pathfinder was decommissioned in 2017.

The ESA's estimated launch date for LISA is 2037. Until then we'll have to exploit our access to LIGO and its sister observatories around the world to advance our new way of getting information about the universe.

# 1 2

## PULSARS

———

### Silently we went round and round

*Oscar Wilde,* The Ballad of Reading Gaol

I n chapter 4 we introduced the notion that the life of a star can be thought of as a successive set of strategies called upon to counter the relentless inward pull of gravity. The first strategy that all stars use is to initiate fusion reactions to "burn" hydrogen to helium. This is the stage our sun is in right now. Once the hydrogen fuel is gone, the "ashes" of one nuclear fire become fuel for the next. How many nuclear fires a star ignites depends on its mass. For a star like the Sun, the nuclear burning will only go up to carbon.

Once the hydrogen fuel is gone, the number of nuclear fires a star ignites depends on its mass. If the mass of the star is about four times the mass of the Sun, the nuclear burning can go all the way up to iron. But there it stops. Iron is the ultimate nuclear ash—there is no way to get energy from it. This means that large stars, once they have exhausted their nuclear fuel, will start to grow a massive iron core.

As we saw in chapter 4, there are several waystations along the road to the eventual death of the star. For a star like the Sun, the only barrier against gravity that needs to be erected comes from the properties of the star's electrons. Electrons need elbow room, so they can't be compressed beyond a certain point. This creates an outward force (technically, it's called "degeneracy pressure") that balances gravity. This is what creates the Chandrasekhar limit on white dwarves; masses above about that limit will generate enough gravity to overcome electron degeneracy pressure.

The inward pull of gravity in stars with a mass from the Chandrasekhar limit up to about eight times the mass of the Sun will blow right by the electron degeneracy pressure and continue to collapse. At this point some weird things start to happen. For one thing, the outer envelope of the star is blown off in an event we call a supernova. For another, the pressure starts to push electrons into the protons in the iron core, turning them into neutrons.

This core of neutrons continues collapsing until it is typically about eight miles (~12.9 km) across, a size that could easily fit inside the city limits of a modest metropolis. At this point a couple of other forces enter the picture. Like electrons, neutrons need elbow room and therefore create their own degeneracy pressure. In addition, if the mass of the iron core is above 70 percent the mass of the Sun, the strong force between the neutrons adds a little oomph to complement the degeneracy pressure. We call this object a neutron star. We believe that this will be the end state of stars between ten and twenty times as massive as the Sun. Because the star loses so much mass before it collapses, there is a fair amount of uncertainty in this number.

There are a lot of unusual things about this strange beast. Let's go through them one by one.

*Density.* The matter in a neutron star is incredibly dense. On Earth, a piece the size of a sugar cube would weigh more than a mountain, for example. Except possibly for black holes, neutron stars are the densest objects in the universe.

*Rotation.* The standard explanation for the rotation of a neutron star involves calling up a picture of an ice skater going into a spin. At the start, she turns slowly with arms extended. As she pulls her arms in her spin gets faster and faster, until her costume is just a blur. The increase in rotation follows from a principle of Newtonian physics called "the conservation of angular momentum," which holds that the smaller a rotating object is, the faster it will rotate. Apply this law to a star that is condensing into a neutron star, and you see that the neutron star will be spinning very fast indeed. The fastest spinning neutron star ever discovered was seen in 2006. About 28,000 light-years away, it spins at a rate of 716 revolutions per second, about as fast as the blade on your kitchen blender.

*Magnetic field.* One of the properties of magnetic fields that is easy to describe but hard to justify is this: if the magnetic field passes through conducting materials such as a plasma and the material moves, the magnetic field moves along with it. In essence, the field and the material are locked together. This means that when the collapse into a neutron star begins, whatever magnetic field the star has collapses down as well. This causes the magnetic field, now confined to a smaller area, to become much stronger than it was at the start of the collapse. A typical neutron star has a magnetic field a trillion times as strong as the one at the surface of the earth, and the field in some neutron stars, called magnetars, can be a thousand times stronger than that.

This short list surely justifies our claim that neutron stars are some of the strangest objects in the universe. They represent the final strategy a star can employ to fight off the inward pull of gravity. If, however, the mass of the iron core is greater than about twice the mass of the Sun, even the neutrons won't be able to prevent the final victory of gravity, and the star will collapse into a black hole.

The exact process by which happens is still a subject of debate among astronomers, but we can point out one interesting fact: the heaviest neutron star discovered to date weighs in at about 2.5 solar masses, while the lightest black hole comes in at about 5 solar masses. Somewhere between these two, then, is the boundary between staying in the universe as a neutron star and leaving it as a black hole.

## PULSARS

It really didn't look much like a modern radio telescope, even though it made a very important discovery. There were no arrays of big dishes pointed at the sky, no massive computer displays. Instead, there were rows upon rows of wires—120 miles (193 km) of wire in all—suspended a few feet above the ground, waiting to record radio signals coming from the recently discovered objects called quasars (see chapter 10).

Located just outside Cambridge, England, this was the beginning of the Mullard Observatory, now a major installation containing many more conventional telescopes. Sitting in the middle of this array on a chilly November day in 1967 was a twenty-four-year-old graduate student named Jocelyn Bell. In typical graduate student fashion, she had spent a lot of time building the telescope, apparently becoming quite proficient in the use of the soldering iron and sledgehammer. She really knew the instrument, so when a strange radio signal kept showing up she paid attention.

It was a blast of radio waves coming in regularly every 1.3 seconds. With her thesis advisor, Antony Hewish, she began the most important task that experimental and observational scientists have to complete: making sure the signal isn't a spurious mistake due to faulty wiring or the like. This imperative is encapsulated in still another quote from the late physicist Richard Feynman: "The first principle is that you must not fool yourself, and you are the easiest one to fool."

So they checked the wiring and looked for the signal on another telescope, but it just wouldn't go away. They wound up giving it a humorous name: the LGM signal, for "Little Green Men," a reference to the notion that Bell might have seen the first signal from an extraterrestrial civilization. Given the data they had at the beginning, they couldn't rule out that possibility, though they were able to abandon it a few months later when they detected the same sorts of signals in other parts of the sky. Because of the regularity of the radio pulses, these strange new objects were christened "pulsars."

Scientists quickly identified pulsars are rotating neutron stars. To understand how a rotating neutron star could produce the kind of signals Bell saw, think about the structure of the earth for a moment. The planet rotates around an axis that goes from the North Pole to the South Pole. In addition, the planet acts like giant magnet, which also has north and south poles. These poles, however, are not the same as the poles of Earth's axis. The North Magnetic Pole is in Greenland, and the South Magnetic Pole is in Antarctica. When the planet spins around its axis of rotation, then, the magnetic poles move through a circle rather than being stationary like the rotational poles.

A pulsar has basically the same sort of structure as Earth; the planet doesn't rotate nearly as fast or have as strong a magnetic field, but the motion of the magnetic poles is the same. Intense magnetic fields in a rotating neutron star move particles toward the magnetic poles. As the particles jet out along magnetic field lines, they begin to emit intense beams of electromagnetic radiation, like beams from a lighthouse that swing around in a circle driven by the star's rotation. If the path of that beam happened to cross the earth, then a terrestrial observer would see a flash or pulse when the beam swept by followed by a period of darkness until the beam came around again. This is exactly what Bell's telescope saw.

Every pulsar is a neutron star, but not every neutron star is a pulsar. Even neutron stars with jets and beams need not be oriented in a way that makes them visible from Earth. Theorists have suggested that there may be as many as a billion neutron stars scattered around the Milky Way galaxy, for example, but only about two thousand have been identified as pulsars.

Before we leave the story of the discovery of pulsars, we have to mention an event that still sparks intense debate in the scientific community. In 1974, the Nobel Prize in physics was awarded for the discovery of pulsars. The debate centers on the fact that the prize was shared by Antony Hewish (1924–2021), Bell's thesis advisor, and astronomer Martin Ryle (1918–84). No one denies that these men deserve the prize. The debate centers on why Bell—now Dame Jocelyn Bell Burnell— was not included. Was it simply antifemale prejudice? Was it the fact that she was a student? Bell herself has never voiced a complaint, and she has been awarded many honors since then, including the $3 million Breakthrough Prize.

## THE HULSE-TAYLOR BINARY PULSAR

In 1974 two astronomers at the University of Massachusetts at Amherst discovered a very unusual pulsar system. Joseph Taylor was a professor, and Russell Hulse was his PhD student. Working with the great radio dish at Arecibo in Puerto Rico, the two began observing a pulsar that at first seemed quite ordinary, if a bit fast. The pulsar rotated on its axis seventeen times per second, producing radio pulses every fifty-nine milliseconds. They noticed, however, that overlain on the standard pulsar signal was a systematic variation in the arrival time of the pulses, a variation that seemed to repeat every 7.75 hours. They realized that this complex set of data could be explained if what they were seeing was a binary system in which one of the members of the binary was a pulsar. (It was only later that the other member was identified as a neutron star.)

Binary star systems are quite common; most of the stars you see in the sky are actually two-star systems, and fully 50 percent of the stars in the Milky Way galaxy are estimated to be binary. Some of these have to be dual systems in which both partners are massive enough to evolve into neutron stars, and it is now obvious that Taylor and Hulse had found one of those.

Subsequent work established that the two neutron stars weigh in at about 1.4 solar masses each. No radio emissions have ever been detected from the companion star, which means that it either is not emitting radio waves or that the radio jets it is emitting do not come in the direction of our telescopes.

The primary importance of the Hulse-Taylor binary pulsar system is that it provides an ideal laboratory for tests of general relativity. This is important, because the effects of general relativity are typically very small and hard to measure, given again that gravity is a very weak force. For example, the perihelion shift of Mercury attributable to general relativity, that is, the amount the distance of closest approach to the Sun changes over time, is forty-three seconds of arc per century. In the binary pulsar, of the other hand, the masses are so great that the periastron, the distance of closest approach of the stars, varies by about four *degrees* per year.

The most important feature of the orbital motion, however, came from tracking the loss of energy over time. In chapter 11 we introduced the concept of gravitational waves as one of the definite predictions of general relativity. At the time of the discovery of the binary pulsar there was no way to confirm Einstein's prediction of the existence of these waves, so for decades the orbit of the binary pulsar was the only quantity we could measure that gave us evidence, however indirect, that gravitational waves existed.

Here's how the indirect evidence worked: according to general relativity the orbiting neutron stars should be emitting gravitational waves carrying about 2 percent of the amount of energy emitted by the Sun in the form of visible light. This energy has to come from somewhere, and the only available source is the orbital motion of the pair. The orbital period that supplies this energy decreases by a little over seventy microseconds per year, corresponding to a decrease in the distance between the stars of about 3.5 meters (~11.5 ft.). It will be about 300 million years before this decrease brings the stars close enough together to collide.

In 1993 Hulse and Taylor were awarded the Nobel Prize in physics for their discovery.

# 13

## THE COSMIC HUM

—

Heard melodies are sweet, but those unheard
Are sweeter

*John Keats,* "Ode on a Grecian Urn"

**S**cience magazine, published by the American Association for the Advancement of Science (AAAS), is one of the most prestigious scientific publications in the world. Getting an article published in this journal is considered a great honor for any researcher. Every year the editors of the journal pick one scientific advance and call it the "Breakthrough of the Year." In 2015 the title went to the detection of gravitational waves (see chapter 11). In 2023 the title went to new drugs that had been developed to combat the global obesity pandemic. Among the runners-up, though, was the detection of what the editors called a "cosmic hum," a universal background of low-frequency gravitational waves that was most likely generated by the collision between supermassive black holes.

Galaxies often collide with each other. If, as we believe, most galaxies have supermassive black holes at their centers, the supermassive black holes in colliding galaxies, responding to the strong gravitational attraction between them, will go into a kind of death spiral, eventually culminating in a merger, a process that our theories predict will take about 25 million years. Presumably this process has been going on for billions of years, so a lot of gravitation waves generated by the inspiral and collision must still be sloshing around the universe. This process is what we believe generated the "hum." We should also point out that the collision between black holes is one method that has been proposed for creating supermassives in the first place, which means that the gravitational waves present in the "hum" may have played an important role in the evolution of the universe.

Before we go on to describe the sophisticated technique that was developed to detect the hum, we should talk a little about the properties of waves in general. Imagine that you are at your favorite beach, watching the surf come in. There will be an obvious regularity, that is, waves will hit the beach at regular intervals. Suppose the waves come in every ten seconds. The frequency of the wave is defined to be the number of crests of the wave that go by a fixed point every second. For our example, in which crests arrive every ten seconds, the frequency would be 0.1 Hz, corresponding to one wave crest each second.

The human hearing system detects sound waves in the range of about 20–20,000 Hz. Because the LIGO detector produced the "chirp" we discussed in chapter 11, its sensitivity can't be discussed in term of a single frequency, but it detects gravitational waves in roughly the same frequency range as the human ear detects sound waves.

There is an important difference between gravitational waves and sound. Gravitational waves travel at the speed of light ($c$) through the vacuum of space. Sound waves travel at a speed of about 1,000 feet per second (~305 m/sec) through the air. This provides the basis for the old trick of estimating the distance to a lightning flash by counting the seconds between the time you see the flash and the time you hear the thunder and dividing by five to get the distance in miles.

What the difference in velocity between these two types of waves means is that the distance between the crests in gravitational waves— their wavelength—will be much longer than the equivalent distances in sound. For example, organ pipes can produce sounds near the lower end of the human frequency range, and these pipes are typically a dozen feet or so in length. A gravitational wave at the same frequency might have a wavelength measured in millions of miles, roughly the size of our solar system. Similarly, gravitational waves at the upper end of the human frequency range might have wavelengths roughly the size of Earth. These numbers explain why LIGO had to be as large as it is. An antenna the size of LIGO cannot pick up the low frequencies generated by supermassive collisions, but it does very well at detecting collisions involving massive bodies like stellar mass black holes and neutron stars.

## MILLISECOND PULSARS

In chapter 12 we laid out the basic mechanism that produces pulsars. They are small and dense objects, the corpses of giant stars that have exhausted their nuclear fuel, exploded as supernovae, and collapsed down to their present size. This rapidly spinning object typically emits beams of radio waves from its magnetic poles. These radio beams sweep through the skies as the pulsar rotates, acting more or less like the radio version of a lighthouse beam. If that beam sweeps past the earth, we see regularly timed pulses of radio waves (whence the name "pulsar").

Over time, we expect the pulsar's rotation to slow down as the emission of gravitational waves drains away its energy. Against this expectation, it turns out that there are a few pulsars that seem to be rotating at a much faster rate than would be expected from the collapse process alone. These are the "millisecond pulsars," those that rotate with a period of ten milliseconds or less. Because of their rapid spin, millisecond pulsars tend to have a very stable rotation rate, which makes them ideal "clocks" for observations that require precise measurements of time. They do not seem to be affected by events like "starquakes" that cause unpredicted variations in the rotation rate of ordinary pulsars.

These millisecond pulsars are found in double star systems where, presumably, one star has gone through its life cycle and has collapsed into a pulsar while the other is still functioning as normal. In a situation like this the pulsar will begin pulling material off its larger partner, material that will go into a spinning accretion disk around the pulsar. As this spinning material falls onto the pulsar, it will cause its rotation rate to increase. In the jargon of astronomers, we say that the pulsar is "spun up."

The connection between millisecond pulsars and the gravitational wave hum is easy to state, but very difficult to turn into an actual measurement In essence, many devices intended to detect gravitational waves do so by sensing the change in length of some structure, a change produced by a passing gravitational wave. For example, in LIGO, what is measured is the change in length of the distance between the laser source and the mirror nearly 2.5 miles (4 km) away. In a system using millisecond pulsars, the "arm" of the detection apparatus extends from the radio telescope on Earth to the pulsar, a distance typically measured in thousands of light-years.

As gravitational waves pass by a pulsar, intersecting the line of sight from Earth, the wave causes a small change in the shape of the space-time grid. This in turn causes a small change in the length of the "arm"—a change, in other words, in the distance radio waves have to travel to get to Earth. This change is seen on Earth as a small change in the frequency of rotation of the pulsar.

Depending on the details of the interaction, the spin rate may go up or down. To give one example, if a gravitational wave with a one-year period goes by a pulsar in the arrangement we have outlined, the rotation rate of the pulsar may be seen to be twenty nanoseconds early in July and twenty nanoseconds late in January. A nanosecond is a billionth of a second—the time it takes a light beam to travel about a foot, a very short time compared to the rotation rate of the pulsar. The millisecond pulsar is as good a timepiece as our best atomic clocks, which means that nanosecond-long changes in frequency of rotation can be detected.

Thus, the stable rotation rate of the millisecond pulsars gives us, in principle, a way of detecting the kind of very low frequency gravitational waves that are predicted to be produced in supermassive interactions and collisions. Turning this possibility into a reality in this scheme turns out to be difficult, not least because the predicted waves have incredibly long wavelengths. It is not at all unusual, for example, to see predictions for the lengths of these wavelengths running into dozens of light-years. Even though such a wave travels at the speed of light, it could easily take many years to pass by a terrestrial telescope. Monitoring the slow rise and fall of a wave over this sort of time period is a difficult job. At the very least it requires observing the pulsar for many years, both to establish its normal rotation rate and to document changes due to gravitational waves.

In addition, we know about what is happening to the pulsar only when we receive radio waves from it. As we have pointed out, this means that that information has to traverse a galaxy full of things that can distort the signal before that signal gets to us. Understanding what these systems are and what sorts of distortions they can introduce is a fundamental task of all astronomy that depends on electromagnetic radiation.

Imagine that you are an astronomer who has just been handed new data acquired from a millisecond pulsar and has been asked to say whether it shows evidence for the presence of low frequency gravitational waves. What would you have to do? In essence, evidence for a gravitational wave would show up as a slight increase or decrease in the pulsar rotation rate—a change from the pulsar's normal rate of rotation. But what is that normal rate? To answer this question, you would have to look at the two possible types of change mentioned earlier—those associated with the pulsar, and those associated with the interstellar medium through which the pulsar radio signals come to our telescopes. To deal with the first of these, what you would probably have to do is to watch the pulsar for a long period of time, giving the extraneous stuff time to average itself out.

In practice, the best way to deal with these kinds of problems is to monitor as many millisecond pulsars as you can. You can think of each pulsar as being part of a separate LIGO type system with an "arm" as long as the distance between Earth and the pulsar being observed. In effect, Earth and each millisecond pulsar are the two ends of a gravitational wave detector. General relativity tells us what the gravitational waves from the collisions between two supermassive black holes should look like. We expect that over the lifetime of the universe there will be many of these events, and hence lots of long wavelength gravitational waves.

These predicted gravitational waves are different from the waves seen by LIGO. For one thing, as we have stressed, they have a much lower frequency and a much longer wavelength. Furthermore, unlike the waves seen by LIGO, we cannot associate a given low-frequency wave with a specific event. For example, right now we cannot get enough detailed information about the wave to allow us to specify the masses of the colliding objects.

The technique that has been developed to monitor waves generated by supermassive black hole collisions is called a Pulsar Timing Array. The predictions of general relativity tell us that we should detect a random background of these waves. Furthermore, they tell us that if we look at the correlation of changes in the rotation rates from two separate pulsars we should see a specific pattern that depends on the between the lines of sight from Earth to the two pulsars. This means that each pulsar we observe gives us one more detection of the gravitational wave background.

Obviously, the more pulsars you can monitor in this scheme, the better your chances of success. Today a major international collaboration links telescopes all around the world. The best way to describe this system is to call it a "collaboration of collaborations," its members being the European Pulsar Timing Array, the North American Nanohertz Observatory for Gravitational Waves, the Indian Pulsar Timing Array Project, and the Parkes Pulsar Timing Array in Australia. Aside from the obvious advantage of having a lot of scientific manpower, this network can monitor the entire sky over both the Southern Hemisphere and Northern Hemisphere. In addition, by sharing data, members of the group can build up the long timeline of observations that are needed for detection.

The Pulsar Timing Array does not detect individual black hole mergers, but measures the cumulative effect of many such mergers. We do not ask which atoms emitted a given microwave. Instead, we ask that all the microwaves, taken together, show the expected pattern for a cold universe. In the same way, a Pulsar Timing Array finding does not tell us what supermassive black holes interacted with each other but contributes to a general gravitation wave background—what the editors of *Science* called a "hum." In 2023 the international collaboration announced the discovery of this hum, the echo of supermassive black holes colliding since the earliest times of the universe.

But we do have to know how to connect the output of different pulsars to gravitational waves. When pulsar timing first came to astronomers' attention, two American physicists, Ron Hellings and George Downs, tackled this problem. Using the Einstein equations, they found in 1984 that if you looked at pairs of pulsars and calculated the change in timing that would result from a general background of long-wavelength gravitational waves and matched it against the angular separation of the two pulsars, you would get a characteristic curve. When the angles between the pulsars is small, the waves tend to be correlated to each other, while when the angle approaches 90 degrees, there is an anticorrelation. Now called the Hellings-Downs curve, this then becomes a way of making sure that the variations in timing that you see are in fact due to a "hum" of gravitational waves. In 2023, for example, the North American Observatory for Gravitational Waves published their analysis of over fifteen years of pulsar timing data. From the sixty-seven pulsars they observed, they created no fewer than 2,211 pairs of data points. When their data fell almost exactly on the Hellings-Downs line, they felt confident in claiming that they had strong evidence for the existence of a gravitational wave background.

That this long-wavelength background is due to the interaction of supermassive black holes is, as of this writing, the best explanation we have for the data now in our possession. We still aren't able to rule out definitively several other alternative sources for the observed changes in pulsar rotation rates. To take one example, some theorists have suggested that at least part of the background by cosmic strings, one-dimensional objects that theoretically form at boundaries of regions of the universe where different sequences of cosmic evolution are found. A cosmic string could certainly, in principle, affect the rotation rate of a pulsar—if it actually existed.

Despite this uncertainty, we believe that the best explanation for the observed regularities in the rotation rate of the hundred pulsars monitored by the collaborating organizations is that they are evidence for the inspiraling and collisions of supermassive black holes. It is certainly an explanation that fits well with other things we know about the evolution of galaxies.

The way scientists deal with this sort of ambiguity is to ask how likely it is that a given result is produced by chance fluctuations. This is generally expressed in terms of a quantity known as "sigma," which gives the probability that the result that was obtained was the result of a chance fluctuation—a fluke. In monitoring a signal from a pulsar, for example, there will always be small fluctuations up and down. This means that there is always a chance that the same fluctuation will show up at some point in two pulsar signals. The sigma number is a way of estimating, statistically, the chance that the coincidence is a real effect or a fluke.

The rigid standard in many fields of physics is that experimental results are questioned until they come from a five-sigma analysis. Typically, this means that the odds of the result coming from this kind of statistical fluke is about 1:100,000. In other words, if you ran the experiment 100,000 times, you would get your result from chance alone just once. The results from the Pulsar Timing Array are currently reckoned to be between three and four sigmas—pretty high, but still not completely up to the accepted standard. Still, it's good enough, we think, to have preliminary faith in the results.

# 14

# DARK MATTER SEARCHES

———

**You're entitled to say, if you're so smart,
why don't you tell me what dark matter is?
And I'll have to confess I don't know.**

*Astrophysicist Jim Peebles*

**A**ccording to our current thinking, something called dark matter played an important role in the formation of the universe as we see it today. While dark matter does not interact with electromagnetic radiation, it makes its presence known through its gravitational effects. We have a robust theory that explains a good deal about how dark matter governed the formation of galaxies and galactic structure. We know that dark matter makes up about a quarter of the mass of the universe. In fact, there is only one important question that we do not know how to answer:

What is it?

Before we get into a description of the so far unsuccessful attempts to detect dark matter in the laboratory, we should probably take a moment to deal with an unfortunate linguistic issue that has crept into modern cosmology. In the late twentieth century astronomers measuring the Hubble expansion made a startling discovery. Instead of slowing down—an effect you might expect to see gravity produce—the expansion is actually speeding up. This was attributed to an unknown material that came to be called dark energy. Dark energy, whatever it is, makes up about 70 percent of the universe. On that note, the world with which we are familiar—the world that scientists have spent millennia examining—constitutes only a bit less than 5 percent of the mass of the universe.

Thus there is dark matter, whose effect is to use the force of gravity to slow down the Hubble expansion, and dark energy, which acts like antigravity and speeds the expansion up. You can think of the two as being like the gas pedal and brakes on your car. Other than the fact that the two have "dark" in their title and the fact that nobody has any idea of what either one is, dark energy and dark matter have nothing in common.

Since our main mission in this book is to explore the world of supermassive black holes, we are going to focus our attention on dark matter, if for no other reason that both dark matter and supermassive black holes seem to be involved in the formation of galaxies. But there is another, somewhat tenuous relationship between dark matter and black holes. Some theorists consider what are called primordial black holes (see chapter 3) to be viable candidates for dark matter itself.

## MAKE, BREAK, AND SHAKE

One author, Jim, characterized the kinds of searches we can do for dark matter by the amusing phrase "make, break, and shake":

- We can try to create dark matter particles in accelerator collisions. There is an ongoing dark matter search at the Large Hadron Collider, for example.

- We can look for specific decay products—breaks—that some theories predict will be produced by certain kinds of dark matter particles.
- We can try to detect the (very rare) events where dark matter collides with ordinary matter, jostling it in a way that allows us to see the collision.

When dark matter was first detected, anything capable of exerting a gravitational force and was difficult (or impossible) to detect was considered grist for the dark matter mill. For example, some theorists suggested that dark matter was actually a large flock of Jupiter-sized objects or, possibly, primordial black holes at the edge of the galaxy. They were called MACHOs (Massive Compact Halo Objects). The name seems to have been proposed as an alternative to the suggestion that dark matter was composed of weakly interacting massive particles (WIMPs). The MACHO moniker has pretty much disappeared, but, as we'll see, WIMPs are still very much around.

In accelerators, particles are brought up to speeds near that of light and allowed to collide. Typically, the region where the interactions take place is surrounded by all sorts of detectors whose job it is to record the results of the collision events. Its job done, the beam is typically cast aside into a large block of metal called, appropriately enough, the beam dump. Early on there is a standard ritual where the beam dump is examined to see if any unexpected particles were created. Nothing was seen at the Large Hadron Collider that might indicate the presence of dark matter in the beam dump.

The search then moved on to the colliding beams themselves. The detectors around the interaction zone record the paths of charged particles. A typical experiment might go like this: Two protons traveling in opposite directions are brought into the interaction area, and their collision produces a spray of particles. The detectors record only the paths of charged particles but can also see the decays of neutral particles. If we look at the detector readout and see five identical particles moving at the same speed to the right but only four such particles moving at the same speed to the left, we are obviously missing something: there should be another particle moving to the left. One possibility is that a chunk of dark matter (which our instruments couldn't see) had been created and was moving to the left, rebalancing the particles.

One problem with accelerator experiments is that dark matter is not the only hypothetical object that could produce this sort of effect. There are many particles besides those associated with dark matter, both known and hypothetical, that would not be seen by the detectors now in operation, and their presence would have to be ruled out as an explanation for the "missing" matter before a claim for the discovery of dark matter could be made. Other possibilities would have to be ruled out before we would be content with the claim for the discovery of dark matter. Despite years of searches at the Large Hadron Collider in Geneva, Switzerland, the world's most powerful particle accelerator, no situation even remotely similar to this example has been seen. The search will go on, and the discovery of dark matter is just about certain to generate a Nobel Prize for whoever pulls it off.

The second category of dark matter searches focuses on evidence for interactions involving specific though usually hypothetical elementary particles whose existence is suggested by some theoretical calculation. The leading candidate for this sort of search is an as yet undiscovered particle known as an axion.

We have a (so far) extremely successful theoretical explanation of all the known elementary particles. It goes by the prosaic name of the "standard model." This model does a good job of explaining the particles we know about. There are, however, many questions that the standard model cannot answer. Some of these questions involve various types of symmetries we see in the way that elementary particles interact. To give just one example, the model correctly predicts that for every elementary particle we discover there must also be an antiparticle. For example, the electron is a small particle with a given mass and a negative electrical charge. The antiparticle for the electrons (it's called a positron) has exactly the same mass but a positive charge. The subatomic world, in other words, seems to have a balance between matter and antimatter. The problem is that when we look at the universe, we see structures like the solar system made entirely of matter, and nowhere do we see structures made of antimatter.

Why?

This situation is a golden invitation for theorists to think about physics beyond the standard model. The axion, an extremely light particle with no electrical charge, was one attempt to solve this and other outstanding problems in particle physics. For many theorists it has replaced WIMPS as the object waiting to be discovered in dark matter searches. Unfortunately, neither the axion nor any of its hypothetical kindred particles have been seen either in accelerator or cosmic ray searches.

According to theory, an axion left to itself will decay into two photons. Unfortunately, this theoretical decay process may take a time longer than the lifetime of the universe. Theorists have suggested, however, that the process can be accelerated if the axion is placed in a strong magnetic field. At the University of Washington, a resonant chamber is located in the field of a strong superconducting magnet. The detection apparatus is called ADMX (for Axion Dark Matter eXperiment). The experimenters start by making a guess at the mass of the axion, then tune the chamber to absorb the energy that would be emitted by the decay of an axion of that mass. The amount of energy from the decay is tiny, measured in a unit called the "yoctowatt." If they don't see this energy deposition, they move up to the next mass in their search field and repeat the experiment. In this step-by-step process the experimenters will eventually sweep through all the axion masses accessible to their apparatus. As with all the other searches we are describing, at the moment no axions have shown up in ADMXs.

In any case, it is the final category of dark matter searches that has generated the most interest in dark matter searches. Essentially, they are massive searches for those WIMPs, modern updates of the Michelson-Morley experiment—the one that destroyed the idea of an ether that defined a correct frame of reference and, as a side issue, opened the way for the theory of relativity (see chapter 2). The basic idea of these experiments is that if the galaxy is really full of dark matter, as the astronomical data suggests, then Earth's motion around the Sun and the Sun's motion around the galactic center should produce a "dark matter wind" at Earth's surface. Dark matter particles, in other words, should be flowing through your body as you read these words. They probably aren't interacting with any of your molecules, though—remember how weak a force gravity is. Perhaps a few of your atoms will be jostled a bit during your lifetime, but that's not something you would notice. This category of dark matter searches assumes that these rare events actually happen and sets up an apparatus to record them.

It's important to realize that there is an important but seldom stated assumption built into this type of dark matter search. The assumption is that whatever dark matter is, it is something on the scale of elementary articles. It need not be a particle known to theorists, but the apparatus we're about to describe would not detect something like a free Jupiter or a primordial black hole. The real problem in designing a dark matter search is the fact that whatever dark matter particles are, they don't often interact with ordinary matter. This means that experimenters must deal with a low counting rate, and this in turn means they have to be careful about background events, by which we mean anything in the environment that can produce an event that can be interpreted as the sort of event your apparatus is built to detect. For example, suppose you have an instrument that is supposed to detect charged particles produced by a specific interaction. If a random cosmic ray hits that apparatus, it could trigger the detector and record the cosmic ray as a real event.

The experimenter has to be aware this sort of error. One technique to deal with it is to put another detector above your experiment to record the presence of the cosmic ray. You can then "veto" the event, recognizing that the apparatus didn't record the desired event but the arrival of the cosmic ray. Alternatively, you can put your apparatus somewhere out of the reach of cosmic rays. This has become, in fact, the most popular way of avoiding cosmic ray background in dark matter searches.

All of this brings us to the Black Hills of South Dakota, the site of one of America's most important and most unusual scientific research facilities, a mile straight down from the little town of Lead, in a cavern shielded from cosmic rays by a layer of solid rock almost a mile thick. The story of that cavern begins a long time ago, in 1877, when a mining engineer named George Hearst (1820–91) and some colleagues bought a gold mine near Lead. This Homestake gold mine remained productive until it was shut down in 2002, producing, in that period, more gold than any other mine in the western United States and making Hearst and his descendants very wealthy indeed.

Over the years, miners followed the Homestake gold seams down into the ground, carving out a network of tunnels and caverns as they went. By the time the mine had come to the end of its useful life, these tunnels had reached a depth of eight thousand feet (2,438 m). And, as you would expect in any operation as complex and long lasting as his one, there was a lot of heavy equipment being used, and to avoid having to bring that equipment repeatedly up to the surface, occasional large chambers were created where equipment could be stored, such as an auditorium-sized chamber next to one of the main mineshafts—a chamber big enough to house a major physics experiment.

The first experiment, run by Ray Davis of Brookhaven National Laboratories (1914–2006), involved bringing in a huge tank, filling it with carbon tetrachloride, an ordinary cleaning fluid, and monitoring reactions produced by neutrinos from the Sun. Neutrinos are massless particles that interact very weakly with ordinary matter—there are billions of them passing through your body right now, for example—so the shielding from cosmic rays supplied by the rock overburden was very important. The experiment brought to light some very important information about elementary particles, and Davis shared the Nobel Prize in physics in 2002. From the point of view of dark matter searches, though, the important thing was that it established the usefulness of deep mines as places for physics experiments.

The main apparatus used in today's search is an upright cylinder about the size of an SUV. This cylinder is filled with ten tons of liquid xenon. (This experiment, along with others in Japan and Italy, consumed an appreciable fraction of the world's supply of xenon, a trace gas in Earth's atmosphere.) The tank holding the xenon has layers of water and scintillating fluid that allow the experimenter to veto events initiated outside the chamber, as described earlier. This is important, because even though the apparatus is shielded by the rock overburden, that rock itself contains radioactive nuclei whose decay can masquerade as a dark matter events. Even the radioactive nuclei in the experimenters' bodies can cause such background events—a fact that gives us a real appreciation of the difficulty of running an experiment like this.

If a dark matter particle actually collides with a xenon atom, two things will happen. There will be a flash of light that will be detected by instruments outside of the tanks, and an electron that has been shaken loose from the xenon nucleus will be pushed upward to instruments at the top of the tank. From the detection of these two events, experimenters can reconstruct the collision event and verify that it took place inside the tank. Such a detection would be the first evidence of dark matter WIMPs.

This experiment, which now bears the acronym LUX-ZEPLIN, is not the only way of detecting dark matter particles. More than 1,850 miles (2,900 km) to the east of the Homestake mine is another mine, one with a distinctly different history. Located near the town of Sudbury, Ontario, it has been turned into a science laboratory, but a lab with a totally different history. Hundreds of millions of years ago, a large asteroid plowed into the ground, depositing huge amounts of iron and nickel. In the nineteenth century, mining engineers discovered the site of the impact and saw that it contained a great deal of nickel. As was the case with the gold seam at Homestake, the concentration of nickel at Sudbury triggered a major deep mining operation. In fact, if you spent an American five-cent piece before the 1980s, you have probably handled a bit of material from that asteroid that hit Earth so long ago.

Unlike the Homestake mine, the Sudbury mine is still operating, bringing out nickel and copper ores, among other products. It is also the home of another deep underground laboratory some 6,500 feet (~1,980 m) below the surface, dubbed SNOLAB (for Solar Neutrino Observatory) because, like the Homestake facility, it was first used to detect neutrinos emitted by the Sun. It is now the home of another dark matter search.

Like LUX-ZEPLIN, the SNOLAB apparatus takes advantage of its location to use the rock overburden to shield its detectors from cosmic rays. Unlike the Black Hills lab, however, it does not use liquid xenon as a detector. Instead, it uses a rack of ultrapure germanium crystals. The idea is that if a dark matter particle interacts with a germanium atom, it will create a microscopic vibration that will travel through the crystal lattice. When it reaches the outer edge of the crystal, it can be detected by ordinary equipment. Like LUX-ZEPLIN, SNOLAB has run for many years without finding evidence for dark matter particles.

Although the two laboratories we've described may seem to be in competition, in fact they are not. The difference in the way the two major labs detect the interaction of dark matter particles with ordinary matter means that they are looking for different types of dark matter. SNOLAB, for example, will be sensitive to the lightest WIMPs, from a fraction of the mass of the proton up to about ten times that mass. Heavier WIMPs will be seen (assuming they exist) by LUX-ZEPLIN.

We know from all kinds of astronomical evidence that dark matter exists and makes up about a quarter of the universe. We have three different ways of looking for it, none of which has been successful. There are several possible ways to explain this situation:

• Most of the searches have started from the assumption that dark matter is a type of fundamental particle. This need not be the case.
• Experiments of the type we've described actually set limits on things like the mass of particles. It could be that we just haven't reached those limits yet.
• There are many theoretical ideas floating around but which are a bit too far out to get much theoretical attention. Examples include everything from strange warpings of the space-time grid to having dark matter be a manifestation of a realm of physics well beyond the standard model.

Dark matter may have been incorporated into primeval black holes— essentially pulled out of our universe. We'll return to this idea.

# 15

## BLACK HOLES IN A QUANTUM WORLD

---

**The idea of getting something out of nothing may sound absurd, but absurdity is not the worst allegation made against quantum mechanics.**

*Ahmed Almheiri*, How the Inside of a Black Hole
Is Secretly on the Outside

The twentieth century was not kind to classical physics. It's not that the great triad of mechanics, electricity and magnetism, and thermodynamics was proved wrong; it's that they were seen to be limited. Newton's laws worked well for billiard balls rolling across a table, but crank the speed of those billiard balls up near the speed of light and you need relativity to deal with them. Shrink them down to the size of an atom and quantum mechanics enters the picture. The great triad, in other words, is fine for normal-sized objects moving at normal speeds, but new science was needed when you wanted to go outside of those limits.

We've talked about the results of relativity, both special and general, in previous chapters. We have seen how general relativity gives us our best theory of gravitation. What we haven't seen yet is how the other great advance in the early twentieth century—quantum mechanics—affects the behavior of black holes. As it happens, the effects are profound, but to understand them we're going to have to understand a little about quantum mechanics.

Definitions first. "Quantum" is the Latin word for "heap" or "bundle." "Mechanics" is the old word for the science of motion. Quantum mechanics, then, is the branch of science devoted to the behavior of things that come in bundles—or, in the jargon of physicists, things that are "quantized." The point is that the world inside the atom comes this way. Matter comes in lumps like the proton and electron. The same can be said for quantities like electric charge, energy, angular momentum and so on. This is unlike our familiar world, where things like billiard balls can come in any size and move at any speed.

Three concepts guide us in understanding the effect that quantum mechanics has on black holes: superposition, uncertainty and entanglement.

## UNCERTAINTY

As we pointed out in chapter 5, when you look at something like this book, a complex chain of physical processes comes into play. For our purposes, the most important of these is the fact that light bounces off of the book and comes to your eye. In fact, this is a specific example of something that must happen every time we observe nature. There has to be a probe (light in our example), an interaction (the light bouncing off the book) and a detection (which occurs when the light enters our eye). The key point is that in the Newtonian world, the interaction of the probe with the object being observed does not affect that object. The book does not jiggle around because photons of light are hitting it—the photons are just too small to have any discernible effect.

This isn't the case in the quantum world.

In the quantum world, the only probes that we have are other quantum objects. If you want to see an electron, for example, the only way we can probe it is with something like another electron. The standard analogy is to imagine having a long tunnel and wanting to find out if there is a car somewhere inside. The catch is that the only probe you have at your disposal is another car. You can, of course, send that probe car through the tunnel and listen for a crash. What you can't do, however, is assume that the car that was sitting in the tunnel is the same after the interaction with the probe as it was before. In the quantum world, unlike the Newtonian one, interaction (or, if you prefer, measurement) always changes the object being observed.

This fact is the basis of what is known as the Heisenberg uncertainty principle, formulated by the German physicist Werner Heisenberg (1901–76): it is impossible to measure two specific properties of a quantum system with infinite accuracy at the same time. To understand this principle, we need to pause for moment and think about ways that measurements can be uncertain.

Suppose you were asked to measure the length of a bench. You would probably use a ruler. Note that the markings on the ruler are a fixed distance apart—let's say one-eighth of an inch (0.32 cm). This means that if several people use the ruler to measure the length of the bench, their answers will differ by something on the order of one-eighth of an inch. We customarily represent this result by saying that the length is $L \pm \Delta$, where $L$ is the average measured length of the bench and $\Delta$, the Greek capital letter delta, is customarily used to represent uncertainty. In this example, $\Delta$ is about an eighth of an inch.

The Heisenberg uncertainty principle deals with pairs of uncertainties. For example, it tells us that we cannot measure both the speed (*v*) and position (*x*) of a quantum object with infinite accuracy at the same time. In equation form it reads

$$\Delta x \times \Delta v > h/m$$

Where *h* is a number known as Planck's constant and *m* is the mass of the object.

Note that the equation doesn't say that we can't know the position or velocity with infinite accuracy; it just says that if we know either one with infinite accuracy (zero uncertainty) then the other uncertainty must be infinite, that is, if we know the position exactly we can have no idea of the velocity and vice versa.

The position-velocity version of the uncertainty principle is the best-known of these kinds of relationships, but there are other versions. The one most useful for us tells us that we cannot know with infinite precision both the energy (*E*) of a system and the time (*t*) that the system has that energy. In equation form:

$$\Delta E \times \Delta t > h$$

As we will see in a moment, the energy-time uncertainty principle plays a vital role in our understanding of the fundamental forces of nature and the long-term fate of black holes.

As early as the 1930s, the Japanese physicist Hideki Yukawa (1907–81) noticed something important about the energy-time uncertainty principle. Provided that the uncertainty in time is small enough (that is, provided we don't observe the system for too long), $\Delta E$ can be very large—even as large as $mc^2$, where *m* is the mass of an elementary particle and *c* is a constant (namely, the speed of light). This means that as far as we are concerned, that particle can actually be present provided that it disappears in a time $\Delta t$. These sorts of particles, which quantum mechanics allows to pop up and disappear, are known as virtual particles. Think of them as the particle analogue of Cinderella at the ball: she could be there provided she got back home by midnight.

Actually, the appearance of virtual particles is governed by the usual conservation laws of physics. You can't make a single virtual electron, for example, because this would require the creation of an unbalanced electrical charge. Therefore, we would expect two virtual particles in this case—a negatively charged electron and a positively charged positron (the antiparticle of the electron). This particle/antiparticle pair can pop into existence, and provided they "unpop" fast enough, no laws of physics would be violated.

This situation has profound significance for the idea of the vacuum. The quantum mechanical vacuum is not a dull place where nothing happens. Instead, it is full of virtual particle pairs popping into and out of existence. We can even measure the effect these pairs have on the electric and magnetic properties of ordinary particles who happen to be in the vicinity of a virtual pair, so we can cite experimental evidence for the existence of virtual particles.

More important, however, Yukawa realized that you could have a situation in which two ordinary particles were passing near each other as shown in the diagram. One of those ordinary particles could emit a virtual particle and, provided that the other absorbed it in a short enough time, the energy-time uncertainty relation would guarantee that no laws of physics were violated. Yukawa realized that a particle—even a virtual particle—exchanged in this way would generate a force. Think of two ice skaters throwing a medicine ball back and forth. Every time one skater throws the ball, he will recoil, as will the other skater when she catches the ball. Newton's first law tells us whenever there is a change of motion like this, a force must be acting.

This sort of mechanism is now known to describe three of the four fundamental forces of nature, with the forces differing only in the sort of virtual particle that is exchanged. Those three forces are the strong (which holds quarks together in particles), the weak (which governs some radioactive decays), and the familiar force of electromagnetism. This picture of particle exchange, which gives us what we can call a dynamic picture of the nature of force, leads us to a fundamental question (some would say *the* fundamental question): Why is gravity so different? As we have seen, our modern ideas of gravity depend on the warping of the space-time grid—what we can call a geometrical idea of force.

One of the great quests of modern physics is to get rid of this fundamental difference between gravity and the other forces. This is usually referred to as the search for a quantum theory of gravity or, perhaps more grandly, a theory of everything. To date, no such theory has been found.

## ENTANGLEMENT AND SUPERPOSITION

The existence of virtual particles and the exchange theory of forces have been around for a long time, and physicists have had plenty of time to get used to them. In a sense, the notion of entanglement is just as old, but until quite recently it was regarded as a possible flaw in the whole quantum mechanical picture. This is because in 1935 Albert Einstein and two colleagues, Boris Podolsky and Nathan Rosen, published a paper that called attention what they thought was a fundamental flaw in quantum mechanics. As always, it's easiest to understand Einstein's argument by using a *gedanken* experiment (see chapter 1).

Imagine you have a box with gloves in it, right-hand gloves and left-hand gloves, Furthermore, suppose there was a rule that said that at the start of any experiment equal numbers of right- and left-hand gloves had to be thrown out from the box. Suppose a pair of gloves is emitted and that the emitted gloves travel in opposite directions. After the emitted gloves have traveled a long distance, one of them is measured. If the measurement turns up a right-handed glove, we know that the other glove is left-handed *without ever measuring it.* This is true even if the gloves have separated so far that a light signal could not pass between them when the measurement was made.

Einstein argued that the only way this could happen was if each glove was either right- or left-handed for its entire journey. There must be, in other words, "hidden variables" that determine the state of the emitted gloves as they travel. To understand his argument, we have to go to the third principle mentioned earlier, superposition. The point is that in quantum mechanics things like the position and velocities of particles, and even the particle identity, are described in terms of probabilities. The collection of probabilities that describe a particle is called "wave function." While in flight, the gloves in our example can be thought of being simultaneously in the state of being right-handed and the state of being left-handed. It is said to be in a state of superposition.

In the standard Copenhagen interpretation of quantum mechanics, the measurement itself forces the system into a specific state— either right- or left-handed, in our example. Once one glove has been forced to choose one form of handedness, the other must choose the opposite, even though no signal can pass between the two gloves. Einstein found this "spooky action at a distance" unacceptable and argued that it constituted proof that quantum mechanics was incomplete.

In 1964, the Irish physicist John Bell (1928–90) wrote a paper in which he showed that in experiments like the one described here, with particles like photons being sent out rather than gloves, certain quantities would be different if the particles were described by quantum mechanics than if they were described by a hidden variable theory. The experiments Bell described were extremely difficult, and it was a full decade before the verdict came in: quantum mechanics was clearly the winner.

It was the only time we can think of in which an exciting new theory was tested in state-of-the-art experiments and found to be correct, and, as a result, many scientists were miserable. Bell's work and the subsequent experiments showed that, unlike Newtonian physics, quantum mechanics is not local. Look at it this way: if a billiard ball rolls along a table, it can influence another billiard ball only if it actually collides with that billiard ball. No matter what it does, it can't influence the motion of a billiard ball on a different table, particularly if that billiard ball is on a table in another galaxy (to take an extreme case). This is our familiar world, where Newton's laws prevail, a world where all interactions are local.

It is not, as Bell's paper showed, the world inside the atom. In the typical experiment outlined above, quantum mechanics demands that the left-going object be described by a superposition of states, and the same is true of the right-going object. Neither is in a well-defined state—in our example, neither is totally a left- or right-handed glove. It is only when a measurement is made that one of the gloves is forced to be either right- or left-handed. Then—and here comes the "spooky" part—the other glove becomes the opposite of the one that was measured. This happens *even if* the two gloves are too far apart for a light signal to pass between them. (This was the point of having those two pool tables in different galaxies.)

This is how physicists think about the weird process: when the two particles are together before the process starts, they are described by a single wave function. We say they are "entangled." As they separate, they remain entangled as the wave function changes. When one particle is measured, the wave function changes—the particle being measured is forced into a specific state. It becomes a right- or left-handed glove in our example. The wave function changes, and the other particle becomes right- or left-handed *without any signal passing between the two*. It is the change in the wave function that does the job. It is also the change in the wave function that is the essence of entanglement. It makes the world of the atom nonlocal, and this is a fundamental difference between our everyday Newtonian world and the world described by quantum mechanics.

A word of caution: don't try to visualize how the process of entanglement works. That way lies madness. Just accept that particles that are once close together can continue to influence each other even when they cannot communicate with each other by means that are familiar in our normal Newtonian world.

## HAWKING RADIATION AND THE DEATH OF BLACK HOLES

The late Stephen Hawking was the first person to successfully apply quantum mechanics to the study of black holes. The result of his work can be thought of as comprising two different approaches, both leading to the same result: black holes are not eternal but have finite lifetimes. This in turn has led to a situation involving something called the "information paradox," a deep theoretical problem that we'll describe in a moment.

We'll begin, however, with the simplest of Hawking's results—the presence of so-called Hawking radiation. We have seen that the quantum mechanical vacuum is not simply empty space but a dynamic system in which virtual particle pairs are popping into existence and disappearing all the time. Imagine, then, that a virtual particle pair pops up near the event horizon of a black hole. It is possible that one member of the pair will fall into the black hole, never to be seen again. The other member of the pair, now abandoned, will wander off.

The lone particle requires energy, both to make up its mass (remember $E = mc^2$) and to supply whatever kinetic energy it needs to climb out of the black hole's gravity well. That energy has to come from somewhere, and the only energy source in the region is the mass of the black hole itself. Thus, we have a situation where particles are being ejected from the region near the black hole event horizon, particles that are referred to as "Hawking radiation," and as a result energy is being drained from the black hole.

This means that over time the mass of the black hole will shrink until, eventually, the black hole simply disappears. The standard visualization of this process is to compare the shrinking black hole to a puddle of water left on the sidewalk after a summer storm. Over time the simple process of evaporation will make the puddle shrink and eventually disappear. In the same way, the slow drain of Hawking radiation will shrink the black hole until it, like the puddle, simply disappears.

The "evaporation times" for black holes are incredibly long. It is estimated, for example, that it will take a stellar black hole as much as $10^{60}$ years to vanish. That's a huge number—a one followed by sixty zeroes. (For comparison, the entire lifetime of the universe since the Big Bang is a one followed by "only" ten zeroes.) So although we know that in the end black holes will not survive, the time scales involved for their disappearance are so long as to be virtually unintelligible to anyone but a professional cosmologist. To drive this point home, we note that the estimated lifetime of the kind of supermassive black holes we've been talking about in this book is about $10^{100}$ years.

## BLACK HOLE THERMODYNAMICS

We mentioned that there was another way to come up with finite lifetimes for black holes—a field known as black hole thermodynamics. We give a short discussion of that field here, but readers with a low tolerance for mathematical formalism (a group that includes both of us authors) can skip ahead to the next section without loss.

Consider an inflated tire on your car. There are two ways to think about the air in the tire. One way is to look at large-scale numbers that characterize the air as a whole—pressure, temperature, volume, and so on. This is the approach of a field of science known as thermodynamics, one of the fields that we called the triad earlier. Another way is to note that the air is composed of many molecules and to use quantum mechanics to describe their individual states. The field of science that connects the atomic and bulk interpretations of matter is called statistical mechanics.

Theoreticians like Stephen Hawking noted that there were many possible ways to create black holes, a situation somewhat analogous to the many quantum states available to air molecules in your tire. From this they were able to use complex mathematics to conclude that black holes actually have a temperature.

Several times in our discussions we have encountered the basic law of physics that tells us that every object with a temperature above absolute zero must give off radiation (infrared, in the case of the human body). One of the basic characteristics of a black hole, on the other hand, is that nothing gets out from inside the event horizon. Theorists expected, then, that black holes would be at zero temperature. The fact they aren't means that they must be giving off radiation, and this in turn means that they must evaporate and die. (The calculated black hole temperatures are very small, typically a tiny fraction of a degree above absolute zero.)

## THE INFORMATION PARADOX

The evaporation scenario for black hole gives rise to a serious problem known as the "information paradox." We'll describe the paradox here, then go into the many solutions that have been devised to resolve it in the next chapter.

Let's start with an evaporating puddle of water. As each molecule leaves the puddle, we could in principle follow it on its outward journey. At some point in the future we could use the information about the molecular tracks to reconstruct the original puddle. The information contained in the molecular structure of the puddle, in other words, is not destroyed by the evaporation. In the same way, if you burned or shredded this book (not a recommended procedure), you could in principle regain the information it contains.

Both of these imaginary calculations illustrate an important point: it is a basic law of quantum mechanics that information cannot be destroyed. It can be shifted around, but it can't be lost.

Now consider the evaporation process of a black hole. The creation of Hawking radiation drains the mass of the black hole but does not depend in any way on how the black hole acquired that mass. It makes no difference whether you throw a book or a rock past the event horizon: the mass of the black hole will increase by the same amount, and the Hawking radiation attributed to the increased mass will be the same. This means that when the black hole has evaporated, any information that was contained in objects that fell into it will be lost. Throw this book past the event horizon, in other words, and we will not be able to reconstitute it by measuring the Hawking radiation. This in essence is the information paradox.

# 16

## BLACK HOLES, QUANTUM GRAVITY, AND ALL THAT

───

**There was a door to which I found no key**
**There was a veil past which I could not see**

Rubáiyát of Omar Khayyám

**B**lack holes aren't just some of the strangest beasts in the universe. They also embody what many consider to be the most important frontier in modern science. We can begin to understand why this is so by recalling that by the end of the twentieth century there were two great pillars of physics shaping our view of the universe. One was the Newton-Einstein pillar, epitomized by general relativity. The other was quantum mechanics. Both of these theories have met numerous experimental tests, and each is supreme in its own world. Newtonian mechanics and general relativity describe the world of normal sized objects moving at normal speeds and give us out best theory of gravity. Quantum mechanics, on the other hand, describes the subatomic world, as discussed in chapter 15.

We want to emphasize how totally different these theories are and how different are the circumstances in which they would be applied. You would never dream, for example, of using quantum mechanics to plot the trajectory of a space probe being sent to the outer solar system. Neither would you dream of applying general relativity to the design of a computer chip. The late paleontologist Stephen Jay Gould coined an awkward but very useful phrase—nonoverlapping magisteria—to describe the difference between religion and science. We can borrow that phrase to talk about the two pillars of science. The idea is that each pillar reigns supreme in its own area and has little to do with the other.

This distinction may work for discussions of science and religion, but it goes against the grain for many scientists. Theoretical physicists being what they are, for example, they are unwilling to settle for a permanently divided universe. For reasons we'll go into more fully, some of the best minds in the human race have spent the better part of the last half century trying to develop a theory in which Newton/Einstein and quantum mechanics are seen to be different aspects of a single underlying theory of everything.

This is where black holes come in.

Think for a moment about how we can test a theory like general relativity. Since gravity is such a weak force, tests of general relativity tend to involve extremely precise of gravitational effects generated by relatively small objects—objects that can't generate large gravitational effects.

How that changes with black holes! The intense compression of the black hole's mass generates gravitational fields that demand the use of general relativity. At the same time, that compression forces us to look at matter at a smaller and smaller scale. If we ever are to understand how black holes work, that understanding will have to include quantum mechanics. We're going to have to have a theory that includes both general relativity and quantum mechanics—a quantum theory of gravity. We don't have such a theory now, and, to be honest, we're not sure we will ever have one. Consider this chapter, then, a sort of progress report about where we are in the quest for the ultimate theory.

## FORCES

We can begin our story by talking about the forces that make the universe run. There are many sorts of forces in our everyday lives— contact forces, magnetic forces, gravitational forces, and so on. From the point of view of a scientist, however, all of those forces can be understood as some combination of just four fundamental forces. In order of strength, they are:

- The *strong force* holds elementary particles and atomic nuclei together. It has to be strong because it has to overcome things like the electromagnetic repulsion between protons in a nucleus and quarks in an elementary particle.
- The *electromagnetic force* holds notes to your refrigerator and powers our national electrical grid.
- The *weak force* governs certain kinds of radioactive decays.
- *Gravity* is the familiar force described by general relativity.

The conflict between relativity and quantum mechanics we described is reflected in the fact that the first three of these forces are described by the process of virtual particle exchange we talked about in chapter 15.

This process, let's recall, describes forces as being mediated by the exchange of virtual particles, a process driven by the energy-time version of the Heisenberg uncertainty principle. The theory that is involved in describing the first three forces in the list goes by the prosaic name of "the standard model." The differences between the three forces depends on the differences between the virtual particles being exchanged. For these reasons, we refer to these three forces as "dynamical."

The strong force between quarks, for example, is mediated by the exchange of particles called "gluons" (they "glue" the particles together). This is a very strong but short-range force. The familiar electromagnetic force, in turn, is mediated by a virtual version of a familiar particle—the photon. Because the photon is a massless particle, this is a long-range force. That's one reason we feel its effects in the macroscopic world.

The weak force is mediated by the exchange of heavy particles known as the W and Z bosons, and, as we said, governs certain radio-active decays. Because these particles have a mass significantly higher than the mass of the proton, the weak interaction is a short-range force.

The first three forces in our list, then, are mediated by the exchange of virtual particles, and the differences between those parti-cles account for the differences between those forces. It also means, as we will see, that at very high energies, when the differences between the virtual particles becomes less important, these three forces can be thought of as manifestations of a single unified force.

But what about gravity? As we saw in chapter 2, general relativity describes the gravitational force as being generated by the warping of the space-time matrix by the presence of mass. We call this a "geometri-cal" mechanism, and it is quite obviously different from the processes that generate the other three forces. Reconciling these two competing views of how forces are generated is what we are calling the problem of developing a quantum theory of gravity.

Before we look at some of the ways that theorists have tried to develop a quantum theory of gravity, let's look at one aspect of the dynamic picture of the generation of forces that plays an important role in modern cosmology—the unification of forces at high energy.

## UNIFICATION OF FORCES

We can begin with a simple analogy for the dynamic generation of forces. Suppose you have two ice skaters starting off at opposite ends of an outdoor ice rink. As the skaters approach each other, one of them throws a heavy object (that medicine ball, say) to the other. When one skater throws the ball, he or she will recoil, and the other skater will recoil when he or she catches it. Suppose further that we can see the skaters but not the object being exchanged. We would see both skaters change their direction of motion. From Newton's first law, we would say that a force had acted. If we were now allowed to see the object being exchanged, we would realize that that force had been mediated by an exchange mechanism. We would agree that a "dynamical" force had been acting.

Now let's make the situation a little more complicated. Let's have two sets of skaters, and let's assume our ice rink is in Nome, Alaska, in the middle of winter. Assume one set of skaters is going to exchange a block of ice but that the other is going to exchange a bucket of water laced with antifreeze. We would say that there were two different forces acting—one mediated by the exchange of a liquid, the other by the exchange of a solid.

Now let's repeat the experiment in the middle of the summer. Since the temperature is above freezing, we would now see only one force acting—the one mediated by the exchange of a liquid. We would recognize that the fact that we saw two forces acting in the winter was an artifact of the low temperature and not a property of the forces themselves. We say that the two forces have become unified.

In the same way, if we go to high enough energies, we expect to see the dynamical forces begin to unify. The unification of the electromagnetic and weak forces has already been observed at particle accelerators like the Large Hadron Collider in Geneva. Our cosmological theories tell us that the temperature of the universe was high enough to produce this unification when the universe was about ten microseconds old. Before this time there were only three forces—strong, gravitational, and the unified electroweak. After this time the universe has the customary four forces we're describing.

Some theories also tell us that another unification occurred when the universe was much younger. Its age was about $10^{-35}$ seconds old (that's a decimal point followed by thirty-four zeroes and a 1). In this unification, the strong force unified with the electroweak. Before this time, then, there were only two forces in the universe—the unified strong-electroweak and gravity.

This property of the fundamental forces reveals another deep connection between what we have called the three dynamical forces. In a real sense, they are really only different aspects of the same force, and we see them as separate only because we are observing them at a very low temperature, like our skaters in midwinter Nome. (For reference, the current temperature of the universe is only about 3°C above absolute zero.) The question before us, then, is whether gravity can be fit into this scheme, perhaps through another unification. Calculations suggest that if such an event occurred, it would have been when the universe was about $10^{-43}$ seconds old, a unit referred to as the Planck time.

One way of unifying gravity with the other forces, then, would be to try to find a way to represent gravity as mediated by the exchange of some as yet undiscovered particle. Another is to assume that there is some sort of fundamental shift in the earliest stages of the universe that gives us an entirely new concept about the action of forces at the quantum state. We'll outline examples of both of these approaches, but before we do, a word of caution. Both of these approaches involve extremely complex mathematics, which we will largely ignore, and neither has much in the way of experimental backing. This is scarcely surprising. It's what happens when you decide to explore the limits of human knowledge.

## STRING THEORY

The history of human investigation of the basic structure of the universe can be compared to the process of peeling an onion. No matter how many layers you peel off, there is always another layer to explore.

Investigate materials, for example, and you find that they are made of atoms. Investigate atoms and you find that they are made of elementary particles. Investigate the particles and you find that they are made of quarks and leptons.

What's next?

If you take a "business as usual" approach, the answer to this question is obvious—we have to take one more layer and think about what the quarks and leptons are made of. In essence, this is what string theory is about. The central concept of the theory is that quarks are actually the manifestation of vibrations of tiny objects known as strings. When we say "tiny," we are serious. In most versions of the theory, the strings are only about $10^{-33}$ centimeters long. The basic idea, though, is that the strings can vibrate in many different ways, and each different mode of vibration corresponds to a different kind of particle. In a sense, the quantum strings we're talking about are analogous to the string of a violin. A single string can have many modes of vibration simultaneously, producing both the fundamental note and many overtones.

Various versions of string theory have been around since the 1960s, and little is to be gained from following the "boom and bust" story attached to them. From our point of view there is one extremely important point about many versions of these theories. When all of the complex mathematics is worked out, they predict the existence of a massless particle called a graviton, which is the particle whose exchange mediates the gravitational force. In these versions of the theory, then, the gravitational force is generated by the exchange of gravitons.

To put this statement into historical perspective, think of how scientists at different times would have answered the question, "Why don't we float off into space?" Newton would simply have stated that there was a gravitational force between you and Earth. He didn't know why this force existed (remember his famous saying, "I frame no hypotheses"). Einstein would have said that the presence of Earth's mass warped the space-time grid around you and you were just following your own geodesic. The string theorist would say that a flood of gravitons was being exchanged between you and Earth, and this generated the force that keeps you in place.

That's the good news. Unfortunately, there's a whole lot more bad news. For example, one of the things that theorists really have to avoid is having their equations churn out infinities. In the great spurt of theoretical physics that took place in the mid-twentieth century, theorists learned to deal with the infinities that crop up in calculations involving the electromagnetic force. There is, in other words, a tried and tested way of dealing with recalcitrant infinities in your equations. It goes by the name of "renormalization," and believe us, it would make all our heads hurt to describe how it works.

What's important, however, as far as we are concerned is that the infinities in the equations of string theory can only be dealt with if the space-time in which we are working doesn't have the four familiar dimensions, but ten or eleven, depending on which version of the theory you want to use. In other words, to keep the string theory equations from blowing up, we have to be living in a world with more dimensions than we're used to. What this means is that to retain string theory we have to find a way to have some extra dimensions that for some reason we haven't yet noticed.

The standard technique for doing this is to think about observing a garden hose lying on the lawn. If you are looking at the hose from far away, it will appear to be a single line. To specify a position on the hose, you will need only one number: "The spot is three feet from the end," for example. In other words, the hose will appear to be a one-dimensional object.

Get closer, however, and you realize that you will need to specify two more numbers—right-left and up-down. Viewed close up, in other words, the hose is seen to be a three-dimensional object. The two extra dimensions are invisible from far away and become visible only when we move in close.

In the same way, theorists argue, the extra dimensions needed to make string theory work are invisible at low energies and become obvious only at higher ones. The extra dimensions are said to be "compactified." Compactification, in other words, seems to be a reasonable solution to the extra dimension problem.

The ultimate test of any scientific theory is how well its predictions match up against experimental results. Unfortunately, this is where superstring theories don't do very well. For one thing, the difficulty of the string theory mathematics makes it impossible for many versions of the theory to make any predictions at all. Because of this, many traditional physicists question whether string theory is really physics or "just mathematics."

Furthermore, in the one situation in which a clean comparison to experiment might have been expected, the theory failed (or at least did not come through as expected). This situation arose because many versions of string theory incorporate a property called supersymmetry. Without going into detail, theories with this property predict that we will find an entire constellation of "supersymmetric particles" corresponding to twins of the particles we know about. There should be, for example, a supersymmetric version of the electron called the "selectron," a sproton, and so on. The theory community eagerly awaited the completion of the Large Hadron Collider, which was expected to produce a flood of sparticles.

Alas, 2010 came and went, the LHC returned copious amounts of data, and no sparticles showed up. Either they don't exist or they are too massive to be made at the LHC. Either way, this wasn't good news for string theories. It appears that our attempt to formulate a theory of everything in which gravity becomes a dynamic force like all the others has hit a wall. Short of finding a sparticle or some other unlikely event, it may be wisest to look at other ways of attacking the problem of quantum gravity.

On the other hand, there is little doubt that the majority of theorists working in this area still believe that some version of string theory is still the best route to a theory of everything. In essence, they are voting with their feet (or at least their laptop computers) on the question of how to proceed from this point.

## LOOP QUANTUM GRAVITY

The other main contender in the race to find a way of unifying gravity with quantum mechanics takes an entirely different approach to the problem. The central aspect of interactions in string theories is that they involve the exchange of virtual particles *in* space. In these theories, the space-time continuum is simply the unchanging stage on which the actions of the universe are carried out. The central departure in loop quantum gravity is that space-time goes from being a passive stage to being the central player in the game. If you imagine looking at the universe at finer and finer resolutions, loop quantum gravity predicts that you will reach a point, $10^{-35}$ meters, when space and time stop being smooth and continuous and become what can best be described as "chunky." In the language of physicists, they become quantized.

It's hard to picture what quantized space-time might look like, but try this: imagine that you have a yardstick and you start cutting it in half. First you cut the entire yardstick in half, then one of the halves in half, then one of those haves in half, and so on. (The Greek philosopher Democritus, born about 460 BCE, used this process of progressive division to suggest that matter was composed of atoms, a concept with a long history in physics.) One of two things will happen: either you will keep on cutting forever, or you will come to a point where division is no longer possible. The first option would correspond to an infinitely divisible continuum, basically the Newtonian picture. The second option is what leads us to loop quantum gravity. In this theory, once the pieces of the yardstick get down to $10^{-35}$ meters, space-time becomes granular. The "grains" turn out to be little loops, with the loops connected to each other in what is called a spin network or spin foam. We find it useful to picture the structure of quantized space-time as being analogous to a sheet of chain mail. Like string theories, loop quantum gravity predicts the existence of a graviton whose exchange generates the gravitational force.

There is one aspect of loop quantum gravity that could conceivably provide an experimental test of the quantization of space-time. It turns out that the theories predict that the quantization of space causes different wavelengths of electromagnetic radiation to travel at (very) slightly different speeds. This difference is way too small to be seen in modern accelerators or telescopes, but it is theoretically possible that if we looked at radiation coming from an event on the other side of the universe we would see X-rays arriving before radio waves instead of having them both arriving at the same time. It would, however, require enormous improvements in technology to detect this effect.

These two classes of theories mark the edge of our theoretical investigation into the basic structure of the universe. Each basically starts at one end of the quantum-relativity chasm. String theories try to find a way to formulate general relativity in the particle exchange language of quantum mechanics. Loop quantum gravity, on the one hand, starts with what is basically a geometrical picture and by quantizing space-time tries to unify all the forces in what is basically a geometrical scheme.

Which, if either, of the approaches described here will lead to the unification of quantum mechanics with general relativity is an open question. One possibility that no one likes to think about, but which we don't think can be dismissed, is that there is no way to unify these two theories and that we will just have to get used to living in a world in which the very large and the very small are simply not connected. We hope this isn't true, but we can't rule the possibility out.

# 17

## THE INFORMATION
## PARADOX

———

**A paradox is . . . well, it is but it isn't.
Paradoxically.**

*Severn Darden, "Metaphysics Lecture,"*
The Sound of My Own Voice

S cientists tend to be very good at discovering things and very
bad at naming them. This is especially true of the many so-
called paradoxes scattered around the scientific landscape.
According to the *Oxford English Dictionary*, one definition of a paradox
is that "it is a statement or proposition that . . . leads to a conclusion
that is against sense, logically unacceptable, or self contradictory." The
classical Philosophy 101 paradox involves a barber who announces
that he will shave only every man who does not shave himself. Having
set this situation up, the instructor then asks, "Who shaves the barber?"
If he doesn't shave himself, then he must, but if he does shave himself
he can't. That is a true paradox.

The British logician and mathematician Bertrand Russell (1872–1970) introduced this barber paradox. The formal statement of the paradox, in the language of set theory, is this: "Is the set of all sets that are not members of themselves a member of itself?" We should note that the paradox vanishes if the barber is a woman, a possibility that escaped consideration in the early twentieth century.

The so-called Fermi paradox is an example of something that is called a paradox but actually isn't, something often encountered in discussions of the search for extraterrestrial intelligence (SETI). It goes like this: we expect that intelligent life will develop on many planets in the Milky Way. Despite searches, however, we can find no evidence to support this proposition. To quote physicist Enrico Fermi, who first proposed the problem: "Where is everybody?"

This isn't a paradox at all. It is simply an example of nature not behaving the way we expect it to. There is, for example, nothing illogical about the statement that human beings are the only advanced life form in the galaxy. Many scientists, pointing to various special aspects of Earth's history, have made this argument. You may not like this conclusion, but, as we said, there's nothing illogical about it. It does explain, too, what has come to be called "The Great Silence"—the lack of evidence for extraterrestrial life.

The information paradox falls somewhere between the philosophical barber and Enrico Fermi. The statement that information cannot be destroyed is one of the cornerstones of quantum mechanics, one of the best tested and verified theories we have. The fact that it seems possible that the existence of Hawking radiation could lead to the disappearance of black holes, along with all the information that fell into a black hole during its lifetime, thus poses a serious theoretical problem.

On the other hand, the properties of black holes follow from Einstein's theory of general relativity, another of the best-verified theories we have. The "paradox" arises because it appears that when we put these two theories together we come to a contradiction. The contradiction arises because since the 1970s theoretical physicists have been unable to write down a theory of gravity that both includes quantum mechanics and general relativity and has accumulated experimental confirmation.

Forcing quantum mechanics to give up the idea that information cannot be destroyed would make theorists uncomfortable. This is not, however, a totally new situation for theoretical physicists. Think, for example, of what it might have been like for a classical physicist trying to adjust to relativity and quantum mechanics in the early twentieth century. Uncomfortable, yes; illogical, no.

This means that in this chapter we will be able to tell you what the information paradox is and describe a few of the many attempts that have been made to resolve it. We will not be able to tell you, however, a generally accepted method of doing so. This can make discussion of the paradox quite confusing. As one physicist we talked to put it with respect to the information paradox, "There are so many answers that I've forgotten what the question is."

There is, however, a way of thinking about the information paradox that the authors find easy to reconcile with the quantum world. It is a fundamental property of quantum mechanics that if we are given the state of a system at one time, we can use the basic equations of the theory to predict the state of the system at a later time. Conversely, we can run this process backward—given the state of a system now, we can calculate the state of the system at a previous time. The key point is that there is a one-to-one correspondence between the system as it is now and the system as it used to be: one previous state corresponds to one present state and vice versa.

Starting in the 1960s, theoreticians were able to show that every black hole is characterized by just three numbers: its mass, its electrical charge, and its angular momentum. Physicist John Wheeler (1911–2008) coined the phrase "black holes have no hair" to characterize this situation. (He attributed it to a student of his.) What he meant was that there is no feature on the surface of a black hole that would allow you to identify it other than the three "no hair" parameters. Black holes are simply bald. Note that for the simple Schwarzschild black hole introduced in chapter 3, a black hole with no spin and no electric charge, only one parameter—the mass—determines the state of the black hole.

In essence, what the "no hair" theorem says is that two black holes that have the same values for these three parameters will be identical. In the process of stellar black hole formation, for example, any details that might allow you to tell the difference between two different stars—the abundance of a certain chemical element in the stellar photosphere, for example—simply disappear when the star evolves into a black hole.

It is the combination of the no hair theorem and time reversal in quantum mechanics that gives rise to what is called the "information paradox." The point is that there are many ways of creating black holes with the same "no hair" parameters. Two large stars with different chemical composition, for example, could easily evolve into identical mass black holes. This means that if we use only the "no hair" parameters (which, remember, are the only parameters we can know about a black hole) we cannot "run the clock backward" to determine its initial state. More precisely, we cannot determine which of the many possible initial states gave rise to a specific black hole. This, in essence, is the "paradox." Once that black hole has evaporated, any hope of finding that initial state will be gone, along with all of the information about that state that fell into the black hole during its lifetime.

The situation, then, can be summarized this way: we have two theories, general relativity and quantum mechanics, each well verified experimentally in its own field. Each has its own set of fundamental equations and basic assumptions. When we try to combine these two theories, we find that they are not consistent with each other, a fact that leads to the information paradox.

Most of the attempts to deal with this problem involve manipulating the equations and assumptions to find subsets of each that are consistent—that do not lead to a paradox, in other words. This can be an extremely difficult mathematical operation, but we have to stress that it is strictly mathematics, not physics. We can summarize a few of the attempts to resolve the paradox as follows:

- It may be that when black holes shrink down to quantum size, the quantum theory of gravity will restore the missing information and resolve the paradox. Since we don't have that theory, this is probably more of a wish than a prediction.
- It may be that black holes aren't as "bald" as we think. In some theories there are "soft hairs" in the black hole surface that, by some as yet unknown process, attach the missing information to outgoing Hawking radiation.
- It may be that the conservation of information is not really as fundamental a part of quantum mechanics as we thought. This option would require a complete reformulation of quantum mechanics.
- It may be that the evaporation of black holes isn't complete but stops when they reach some as yet undetermined size. In this case the universe would be full of remnant black holes, each carrying the information that entered the hole when it was larger.
- Some string theories predict that some of the Hawking radiation will be in the form of as yet undiscovered particles called gravitons that are not included in Hawking's original calculation. Perhaps these particles carry the missing information.
- Maybe the information is carried away by gravitational waves (see chapter 11).

- Maybe the information disappears from this universe but reappears in another universe that is part of the multiverse. (We're not kidding.)

We could go on—there are easily a hundred different approaches to resolving the information paradox on record. They all share the characteristic of involving complex mathematics, with the math being done pretty much correctly. This gets us to the central issue in resolving the information paradox: Which, if any, of the many possible (mathematically correct) resolutions is right?

The way these kinds of questions are answered in any scientific field is by appeal to experiment or observation. If you have two or more theories being proposed, you find a place where different theories make different predictions about a specific event or process, then you look to nature to see which prediction was right. The answer might come from experiment, or it might come from observing and measuring some natural process, but in the end it is nature that makes the choice for you.

Here we run into a stone wall, because there is no way we can measure the total information content of a black hole unless we are willing to wait $10^{60}$ years for the black hole to evaporate. Even measuring the Hawking radiation or the black hole temperature is well beyond any foreseeable possibility. We're afraid that the information paradox will be with us for some time to come.

## A NOTE ABOUT THE PRESS

One of the aspects of the information paradox that is obvious to everyone who looks into the literature but is seldom discussed is the role of the science press in shaping the debate. Since Jim Trefil has extensive experience in this area, we should make a few comments about the way that black holes are treated in the media.

The first thing to realize is that the scientists who do the work that forms the basis for an article or a book have absolutely no control over the headlines or book titles. These are formulated by marketing people who have a completely different view of the situation from that of the scientists whose work is being used. Perhaps the most famous example of this effect was a book coauthored by the late Nobel Prize–winning physicist Leon Lederman. The book was about the search for what is now known as the Higgs particle, but the book came out with the title *The God Particle*. This undoubtedly increased sales, but Leon frequently expressed embarrassment about the title, which was apparently concocted without consulting him.

In the same way, every mathematically correct advance connected to the information paradox generates screaming headlines to the effect that a decades-old problem is "about to be solved" when in fact all that has happened is that, at best, another plausible item has been added to the list of attempts to solve the information paradox. It's best to cultivate a healthy sense of skepticism when you read these sorts of articles.

# 18

# THE SINGULARITY

___

**In real time, the universe has a beginning
and an end at singularities that form a
boundary to space-time and at which
the laws of science break down.**

*Stephen Hawking,* A Brief History of Time

T here are few words more troubling to a theoretical physicist
than the word "singularity," and few things less welcome, especially if the singularity comes out of some trusted equations.
For a Schwarzschild black hole, for example, the density of matter is
supposed to become infinite at the very center, a point to which we'll
return in a moment. From a physicist's point of view, a singularity
arises when an equation (or set of equations) that is believed to give an
accurate description of reality suddenly tells you that some measurable
quantity is supposed to be infinite. The slang term for this situation is
that the equation "blows up." What makes this situation so difficult is that

it is logically impossible for any physical quantity to be infinite. Big, yes; infinite, no.

The history of physics is full of infinities that people thought constituted a serious problem only to disappear when examined in a different way. Let's look at a couple of these before we tackle the singularities associated with black holes. First, in the late nineteenth century physicists were working on the problem of explaining the way that light interacts with matter. They would shine light into a box with mirrored sides, heat it up, and measure how the energy in the light shifted around. Unfortunately, the only theoretical tools they had at their disposal for handling the atoms in the walls of the box were pictures of the atoms in which electrons were attached to the atomic nuclei by springs. With this picture of the atom and the other tools at their disposal—their best picture of what the atomic world was like—they got a prediction that the energy in ultraviolet light (the most energetic light they knew about) would become infinite in this sort of experiment. This was called the "ultraviolet catastrophe." (We suppose that had they known about X-rays, they would have called it the X-ray or gamma ray catastrophe.)

The German scientist Max Planck (1858–1947) solved the problem of the ultraviolet catastrophe in 1900, when he laid the foundation for the science of quantum mechanics. Among other things, quantum mechanics gives us a much more accurate description of the atom. (If you don't believe this, how are you going to explain that your smart phone—designed by engineers using the laws of quantum mechanics—actually works?) In this historical example we see that the singularity was actually pointing people toward an entirely new branch of science, one that was developed throughout the twentieth century and plays a major role in our life today.

Second, toward the middle of the twentieth century another kind of infinity began to show up in calculations involving the interaction of high energy photons with charged elementary particles. The field is known as quantum electrodynamics (QED) to distinguish it from the James Clerk Maxwell style of electrodynamics that was done in the nineteenth century. The theoretical bases of the field were quantum mechanics and special relativity, so there seemed little point in trying to develop a new kind of science to solve the infinity problem. Instead, the theorists found clever ways to combine the equations in the theory so that the infinities canceled each other out. In essence, they found a new way to deal with the theoretical base they already had, the technique called renormalization that we mentioned earlier. For this work two American physicists, Richard Feynman (1918–88) and Julian Schwinger (1918–94), and the Japanese physicist Shinichiro Tomonaga (1906–79) shared the Nobel Prize in physics in 1965. As an aside, we note that the development of QED contributed to the creation of the theoretical field called quantum chromodynamics (QCD), which is an integral part of the standard model, our best theory of the ultimate structure of matter. The "chromo" in the title refers to the fact that in the standard model quarks have a quantum number called "color." (We'll skip over the reason for that.)

These examples show that singularities and infinities are nothing to be afraid of. The fact that at first glance even the simple Schwarzschild black hole has singularities hidden inside its event horizon doesn't mean that black holes are somehow impossible. At worst, it may mean that we need to develop a new branch of science, such as quantum gravity (see chapter 16).

A singularity arises when your equations tell you that some real quantity that your equations describe—the density of matter inside the event horizon of a black hole, for example—becomes infinite. Since there are no infinities in the real world, the appearance of an infinity in your equations is a clear signal that you are doing something wrong. You might be misapplying the equation to situations in which is does not apply, you might be using the wrong equation, or you might be overlooking a different way of solving that equation.

Here's a simple example of how a singularity might arise and be dealt with. Suppose you had a stringed instrument like a violin or guitar and you wanted to know the relationship between how hard you plucked the string and how loud the resulting sound was. Suppose further that you found a simple relationship between these two quantities: double the energy you put into plucking the string and you double the sound output, for example. Normally you would express this relationship as an equation, just as Einstein expressed the relation between mass and the curvature of space as a (more complicated) equation.

Now suppose I asked the equation to tell me what would happen if I started plucking the string with more and more energy. The prediction of the equation would be unambiguous: the harder I plucked, the louder the sound. If I kept up this routine, I would get to a singularity of sorts; if I put an infinite amount of energy into plucking I would get an infinite amount of sound out. This would be a situation analogous to the ultraviolet catastrophe described earlier.

We're sure that you've already seen the fallacy in this argument. Long before we get to our contrived "singularity" of an infinite sound, the string will break, and no more sound will come out. The appearance of the singularity comes about because we started with a faulty description of the system we were investigating. This is another way that issues involving a singularity can be dealt with, analogous to the solution of the ultraviolet catastrophe supplied by the development of quantum mechanics.

The issue of singularities and black holes is a bit more complex and does not necessarily lend itself to simple solutions. We can start with a simple Schwarzschild black hole, one that has mass but no rotation and no electrical charge. We have already discussed the fact that once we pass through the event horizon of such a black hole, communication with the outside universe is no longer possible. Consequently, nothing we say about the interior of the Schwarzschild sphere can be verified or tested observationally. Nevertheless, we can continue to apply the equations that worked so well outside of the event horizon to the environment inside the horizon. When we do so, we find that the equations predict a true singularity at the center of the black hole. As the distance to the center goes to zero, the equations tell us that the density of matter becomes infinite. This is a situation that is often described as having "the laws of physics break down," but we prefer to regard it as a warning that we are doing something wrong—the kind of warning we got in our simple example.

The analysis of the black hole singularity is made more difficult by the fact that the equations tell us that any bit of matter that is pulled inside the event horizon must eventually wind up at the center, just at the point where the basic laws of physics are supposed to break down.

There is an easy way to visualize why this must be so. One way of picturing the gravitational force of a black hole is to imagine the gravitational force exerted by the black hole as being a flexible sheet being pulled into the central hole, with the sheet moving faster the closer it is to the center of the hole. When the sheet is moving toward the hole at the speed of light, we would say that we were crossing the event horizon. Inside this point, the sheet will be moving toward the central hole at a speed faster than that of light.

When people think of relativity, they usually come up with a sentence like, "Nothing can move faster than light." This is definitely a result of the theory of relativity, but it needs some interpretation. It is true that according to relativity nothing can move between points in space-time faster than light. The theory says nothing, however, about the speeds at which space-time itself can move. As we note the effect of the expansion of space-time during the period of inflationary expansion, space-time itself can move at any speed, including speeds faster than light. It is only speeds *relative to the speed of space-time* that are restricted.

This distinction is relevant to our discussion of black holes. Inside the event horizon, the "sheet" is moving toward the center with a speed greater than the speed of light. Any mass in the region can only move away from the black hole center at the speed of light relative to space-time. In effect, that mass we're talking about is trying to walk up the down escalator while being restricted to speeds less than the speed of the escalator. It will lose and be sucked into the center of the black hole.

We have noted that a black hole can have only three different parameters: mass, electrical charge, and angular momentum (the thing that keeps a body rotating unless it is acted on by a torque). The Schwarzschild black hole is the simplest possible structure, having mass but no charge or rotation. Schwarzschild was able to solve the equations for his black hole in a couple of months, but it wasn't until 1963 that New Zealand mathematician Roy Kerr solved the Einstein equations for a rotating (but chargeless) black hole. The singularity in that case is considerably more complex than it is in the absence of spin.

This is not just a mathematical exercise, because we have good reason to believe that all of the black holes we see are rotating. To understand why this could be so, think about the ways black holes form. If they are the result of a supernova, the star that produced them would have been spinning, and that spin would be carried by the collapsing core as it formed a black hole. If they are the result of a collision, like the black hole whose birth LIGO witnessed (see chapter 11), the two black holes that eventually coalesced would have been locked in a million-year death spiral, and the angular momentum of that spiral would be present in the newly born black hole. Even if the black hole formed directly from a dust cloud, it is extremely unlikely that the cloud would not rotate as it collapsed. We expect, then, that Kerr's solution will describe most (or even all) of the black holes we see.

It turns out that adding rotation to a Schwarzschild black hole introduces a lot of new behaviors. Let's look at a couple of them individually.

As the rotating black hole accumulates more matter, the centrifugal force associated with the rotation grows. This force is not present in a Schwarzschild black hole. Eventually, instead of the point singularity we saw in the Schwarzschild case, the singularity in a rotating black hole will be stretched into a ring. In addition, the equations tell us that there will be two event horizons in a rotating system. They are called, appropriately enough, the inner and outer event horizons. The outer event horizon is the familiar "point of no return" with which we are familiar. The inner horizon needs a bit more explanation.

It's important to realize that the centrifugal force associated with rotation in a Kerr black hole is an *outward* force, capable of countering the inward force of gravity. The place where the outward centrifugal force is balanced by the inward force of gravity marks the inner event horizon. Essentially, all the matter and radiation that has ever fallen into a rotating black hole will wind up trapped at the inner event horizon. It can't go farther in because of the centrifugal force, and it can't go farther out because of gravity. When science-fiction authors write about black holes collecting the entire history of the universe, this is what they're talking about.

Speaking of science fiction, the transformation of the point singularity into a ring brings up some interesting possibilities, at least as far as mathematics is concerned. An object approaching the ring can actually go through it without encountering a singularity. If it does so, the mathematics allows the possibility that the object will enter a structure known as a wormhole, essentially a funnel made from distorted space-time. At the other end of the wormhole would be something known as a white hole. Just as a black hole pulls matter in, a white hole spews it out. This is where the science fiction trope of people winding up in parallel universes and the like comes from. Needless to say, neither wormholes nor white holes have ever been seen. We mention them here only to show how adding rotation to a simple black hole leads to interesting (if completely fictional) consequences.

Finally, solutions have been found for the other two possible black holes: those with electrical charge, with or without rotation. These are less interesting because an electrically charged black hole would quickly attract charged material from its surroundings and become neutralized. Thus, we do not expect to see electrically charged black holes in the real world.

## DEALING WITH THE SINGULARITY

We suspect that a majority of our colleagues assume that black hole singularities will go the way of the ultraviolet catastrophe and disappear once we have a quantum theory of gravity. This is certainly a possibility. We can already see how such a scenario could play out in quantum loop gravity theories, discussed in chapter 16 as one of the candidates for a quantum theory of gravity.

The central point of these theories is that space-time itself is quantized, which means, in essence, that there is a smallest chunk into which space can be divided. This is usually taken to be something called the Planck length, which is about $10^{-35}$ meters, an almost unimaginably small size.

The importance of the quantization of space-time is that it means that the mathematical process that led to the singularity—essentially a process in which the distance between an object and the singularity goes to zero—can't happen. Instead, the singularity is smeared out over a chunk of space-time the size of the Planck length. Stating this in formal mathematical terms, the standard Schwarzschild singularity arises because we can divide by zero when we calculate density, and this is where the infinite result comes from. The quantization of space-time, on the other hand, means that the smallest number we can divide by is the Planck length. No division by zero means no singularity.

Other theorists have devised more complex schemes to avoid the appearance of singularities. In 2023 Roy Kerr created waves in the astrophysical community by publishing a paper that claimed, on the basis of some highly abstract arguments, that singularities would not appear in rotating black holes.

From our point of view, these kinds of debates, while interesting, don't really tell us much about real black holes. It's important to remember that all of this—singularity, wormholes, Planck length, and so on—involves processes that go on inside the event horizon of a real black hole. Since this is a region we can never explore, even in principle, all of these intriguing concepts can never be tested by observation or experiment. Thus, as interesting as the debate over singularities is, it is destined to go on forever without ever being resolved until we have a real, fully tested quantum theory of gravity.

# 19

## SUPERMASSIVES
## AND THE BIG BANG

———

### Maybe the universe is just one of those things that happens from time to time.

*Edward Tryon,* Is the Universe a Vacuum Fluctuation?

Τ he discovery of supermassives in the earliest galaxies by the James Webb Space Telescope strongly suggests that they play a crucial role in the evolution of the structure of the universe. To bolster this claim, however, we have to have a sense of our current knowledge of that evolution. To give us the background we need, we'll devote this chapter to a step-by-step exploration of what we know (and don't know) about how a universe full of galaxies came to be.

First, what was there before the Big Bang?

This ranks pretty high on our "least favorite questions" list. Unfortunately, if you spend any time trying to explain science to a general audience, as we both do, you're sure to run into someone who asks question. Let us first explain why this question is so annoying and then tell you about some of the ways people have tried to think about the pre–Big Bang universe.

The main problem is that time—indeed, space-time itself—didn't come into existence until the Big Bang. We will implicitly recognize this point by giving the times when various unifications occurred in terms of citing time intervals beginning with the Big Bang. This assumes a system where time zero is the Big Bang itself. Given this, it is reasonable to ask what it means to talk about time before time itself began.

It has become customary, in fact, to compare asking about events before the Big Bang to asking something like, "What's north of the North Pole?" If you think about this question for a while, you realize that asking it makes no sense. Any direction you go from the North Pole is south. Even if you get into a helicopter and go up, all you've done is move yourself into a position above the North Pole—you still aren't north of it. You see, it's not that there's nothing north of the North Pole, it's that there's not *even* nothing north of the North Pole.

In the same way, one way of answering the question about what happened before the Big Bang is to say that it's not that there was nothing before the Big Bang—it's that there wasn't *even* nothing before the Big Bang.

Nonetheless, the advent of quantum mechanics and particularly the uncertainty principle has breathed new life into the question of the origin of the universe. Oxford mathematician and physicist Roger Penrose, who received a Nobel Prize for his work on the theory of black holes, has developed a theory in which the universe goes through a never-ending cycle of expansions (Big Bang) and contractions (Big Crunch). In a scheme like this, the question is not, "What was there before *the* Big Bang?" but "What was there before *this* Big Bang?" That is, in principle, a question that can be answered.

It is at about this stage in the discussion that people begin to worry about something else, essentially to ask how we can get something (the universe) from nothing. In other words, if you assume that the universe before the Big Bang was a perfect vacuum (a reasonable, if Newtonian, assumption), then where did the energy to make all this stuff come from?

Imagine that you have a level field and you dig a hole in the middle of it, putting the material you removed into a pile. If all you could see was what was going on above the ground, this would look miraculous: a pile of dirt would suddenly appear from nowhere, a classic "something from nothing" scenario. If you were shown the hole, however, you would realize that nothing unusual had happened. Someone had just moved dirt from one place to another.

In the same way, theorists who think about these sorts of question routinely point out that gravitation potential energy can be both positive and negative, just as your field can have dirt in piles and holes. In their theories, the energy to make the first matter (a pile) just comes from allowing the gravitational energy to become negative (that is, a hole).

Thus, the questions that ordinarily come up when we talk about the origin of the universe can, in principle, be answered without doing damage to the laws of physics.

## THE QUANTUM MECHANICAL VACUUM

Before we go on to talk about how the structure of the universe evolved, we have to address one point in the preceding discussion. We referred to the notion of a perfect vacuum as "Newtonian." When we imagine a vacuum, we usually think that it's a region of space with very few and ideally no particles. One of the best vacuums in nature is in interstellar space, where we find about one particle per cubic centimeter. (A standard coffee cup would hold about seven hundred particles at this density.) The surface of the moon, which we usually think of as a pretty good vacuum, may have something like eighty thousand atoms per cubic centimeter. The best man-made vacuum is in the beam line at the

Large Hadron Collider in Geneva, Switzerland, and has around a factor of a thousand fewer particles than we find on the moon.

But are these vacuums really empty? In the Newtonian vision of the universe they are, but when we examine them through the lens of quantum mechanics we get an entirely different picture. In chapter 15 we talked about how the energy-time uncertainty principle allows virtual particles to exist for short times. We used this notion to come to an understanding of the quantum mechanical view of forces. The energy-time uncertainty principle also has something to say about the nature of the vacuum.

Consider an expanse of seemingly empty space. The uncertainty principle tells us that a particle/antiparticle pair can pop into existence in that space as long as it disappears in the allowed time. These virtual particle pairs can pop up anywhere in our presumed vacuum, as long as they disappear quickly enough. Observing what we think is a vacuum will be a little like watching popcorn pop. At random times, we will see the "vacuum" producing virtual pairs. The only difference between our quantum mechanical "popcorn" and what you might buy at a movie theater is that the quantum mechanical popcorn has to be able to "unpop." The quantum mechanical vacuum, then, is a dynamic place, full of particles popping into existence and immediately unpopping back out of existence.

In the early days of theoretical work on the origin of the universe, physicists looked at the possibility that the entire universe was the result of a single quantum fluctuation. Today people who want to go the quantum fluctuation route generally take a more modest approach and assume that the fluctuation distorts the space-time grid enough to set an expansion in motion.

The next question is to ask how we got from the beginning to where we are now.

The Planck time is the time when our theories tell us that all of the four fundamental forces became unified. The nascent universe was as simple as it could be, one kind of particle interacting through a single unified force. As we enjoy pointing out to our students, it's been all downhill since then.

According to the standard model, the strong force decouples from the electroweak at $10^{-35}$ seconds. From this point forward the universe is a churning sea of quarks and leptons governed by the action of three fundamental forces: strong, electroweak, and gravity. This is also the time when the universal imbalance between matter and antimatter shows up. Roughly speaking, for every ten billion particles of antimatter being created, there are ten billion and one particles of matter. (The extra particle is what you and I and the rest of the observable universe are made of.) The annihilation process produces a sea of high energy photons that has an enormous effect of the next stages in the evolution of the universe.

## COSMIC INFLATION

During the period between about $10^{-33}$ and $10^{-31}$ seconds, the universe undergoes a period of extraordinary expansion. In a very short time interval, it goes from being smaller than an elementary particle to being about the size of a grapefruit. This period of rapid expansion, first suggested by physicist Alan Guth, is referred to as "inflation." It involves an expansion much faster than the one that Hubble discovered.

There is one aspect of inflation that often bothers people who encounter it for the first time. If you take the initial and final state of the universe seriously and do a standard distance divided by time calculation, you come up with a number much greater than the speed of light. Does this mean that cosmic inflation violates the theory of relativity by allowing speeds greater than the speed of light?

Appearances to the contrary, it does not. What the theory of relativity says is that nothing can travel *through* space at a speed greater than that of light. Cosmic inflation, however, involves the expansion *of* space, not the movement of anything *through* space. Presumably, if you can imagine a microscopic rocket ship inside the inflating universe, the speed of that rocket ship relative to the space-time grid would be perfectly consistent with relativity. What causes the confusion is that we think of observing the rocket ship from outside the universe and adding the velocity of the expansion to the velocity of the ship. Once you realize that by definition there is no "outside" to the universe, the difficulty vanishes.

At ten microseconds, we're now getting into energy ranges reachable by our particle accelerators and the universe starts to become more recognizable. The electromagnetic and weak forces decouple, leaving our familiar four forces. The quarks condense into the ordinary elementary particles with which we are familiar. The universe is a hot, churning plasma full of high energy photons produced by matter-antimatter annihilation.

At three minutes, matter begins to take on a familiar form. Before three minutes, if a proton and a neutron come together to form the nucleus of a deuterium atom, the next collision will knock them apart. (Deuterium is an isotope of hydrogen with one proton and one neutron in its nucleus.) At about three minutes, the temperature of the plasma has fallen to the point that the collisions are sufficiently mild so that nuclei—collections of protons and neutrons—can survive. It is at this point that the nuclei of atoms start to be built up from the primal protons and neutrons in the plasma. The various isotopes of hydrogen, helium, and (a little bit of) lithium—the lightest nuclei in the periodic table—can form.

Less than a minute into this atom building process, however, the Hubble expansion will have carried the particles far enough apart so that no more additions are made to nuclei. The Big Bang thus produces only a few light nuclei—all other nuclei are made much later in stars, as we described in chapter 4.

The universe is now an expanding and cooling ball of particles, nuclei, and electromagnetic radiation. If the matter in the universe tries to come together under the influence of gravity, collisions with the high energy photons will rip it apart—think of a cannonball hitting a stone wall. By the same token, any visible light or other kind of electromagnetic radiation will remain trapped in the plasma, endlessly being absorbed and reemitted by particles and nuclei. In a kind of replay of what happened at three minutes, if an electron in the plasma tried to hook on to a nucleus to form an atom, it will be knocked off in the next collision. Consequently, during this long cooling period, the structure of the universe doesn't change very much—at least the part that we can see.

At 379,000 years, this situation changes. Electrons can now hook onto nuclei and survive. In other words, matter can now take on the form of atoms. This is an extremely important development, because atoms are largely transparent to electromagnetic radiations. The photons that were trapped in the primal plasma are now released and come flooding out. The Hubble expansion stretches out the wavelengths of those photons over the next 13.8 billion years, so that when we look into the sky today we see what we call the cosmic microwave background— messages from the earliest universe carried to us today.

It would be nice if we could give a short summary of the process by which primal matter in the Big Bang condensed under the influence of gravity to give us stars and galaxies and all the rest of the visible universe. Unfortunately, it's not that simple.

## THE COSMIC MICROWAVE BACKGROUND

Let's start with all that electromagnetic radiation that was released when atoms formed 379,0000 years after the Big Bang. In the primal plasma this radiation consisted of high-energy photons: X-rays and gamma rays. Over the next 13.8 billion years, as the universe cooled and expanded, the wavelengths were stretched out until today they are seen as microwaves. The discovery of what is now called the cosmic microwave background (CMB) bears an uncanny similarity to Karl Jansky's development of radio astronomy (see chapter 6).

The 1960s were a busy time for technological advances in communication. Early attempts at long distance television broadcasts were unbelievably primitive by modern standards; they involved bouncing signals from the ground off an inflated balloon in orbit. It's obvious that if you are going to point receivers at the sky to pick up TV signals, you will want to know what else is coming into your receivers. At Bell Laboratories in Holmdel, New Jersey, two physicists, Arno Penzias and Robert Wilson, were given the job of surveying the possible sources of interference with incoming TV signals. They were given an old receiver that was twenty feet (6 m) long and looked something like a first baseman's mitt. It could be swept around to examine all parts of the sky. They quickly learned how to screen out some man-made signals, whether civilian or military, but no matter what they did they found a weak signal that came through as a hiss in their apparatus.

When an experiment produces a constant signal no matter how or where the receiver is pointed, in this case, the first conclusion an experimenter comes to is that there's something wrong with the apparatus. After all, if changing the parameters of an experiment doesn't change the outcome, chances are that there's a wire crossed somewhere.

Penzias and Wilson tried all sorts of schemes to get rid of their hiss. They pointed the receiver at New York City to see if the signal could be blamed on urban interference. They eliminated obvious astronomical sources like the galactic center. In what is arguably the most famous part of their search for an explanation of their hiss, they noted that a group of pigeons had built a nest inside the apparatus, coating the interior with what was delicately described as a "white dielectric substance." Wilson is supposed to have commented later, "We've either discovered the origin of the universe or a pile of bird****". They cleaned out their receiver, but the hiss didn't go away.

Discouraged by their inability to either explain or eliminate the hiss, they began looking at possible theoretical explanations. A friend suggested that they talk to physicist Robert Dicke at Princeton, who was working on consequences of the Big Bang scenario, which was a hot new topic at the time. The Princeton theorists had predicted that at our present era, the universe would be filled with microwave radiation at a temperature of about three degrees above absolute zero.

Bingo. The hiss was real.

What started out as a routine piece of experimental physics turned out to have the deepest possible consequences for our ideas about the universe. At the time there were two viable classes of theories in vogue to explain the structure of the universe—one was the Big Bang, while the other was called the steady state universe. In that theory, the Hubble expansion was attributed to a universal expansion, but as galaxies separated from each other, new galaxies were created by an as yet unexplained process in the resulting void. The advantage of the steady state theory was that the universe was eternal, and we didn't have to worry about how it started. Unfortunately for its proponents, the steady state theory had no way of explaining the cosmic microwave background, so it has been pretty much relegated to the collection of beautiful theories destroyed by an ugly fact.

In 1978 Penzias and Wilson received the Nobel Prize for their discovery.

## THE COSMIC MICROWAVE BACKGROUND AND INFLATION

The CMB contains information about the early stages of the universe, so Penzias and Wilson's experimental findings were quickly followed by a series of satellite observatories designed to examine it in detail. The first of these was the Cosmic Background Explorer (COBE), launched in 1989 and decommissioned in 1993. Unlike the Penzias-Wilson experiment, which looked at only a single frequency of microwaves, COBE examined a wide frequency range and established once and for all that the CMB had the characteristics of radiation emitted by a body at about three degrees above absolute zero.

It is a well-known fact that every object with a temperature above absolute zero gives off a characteristic pattern of electromagnetic radiation. This characteristic pattern of frequencies is known as black body radiation. Right now, for example, your body is giving off radiation concentrated in the infrared. In the same way, the universe is filled with microwave radiation with a characteristic frequency distribution.

The results from COBE were spectacular. They showed the expected pattern of black body radiation with very small error bars. One of us, Jim Trefil, was at the scientific meeting when the COBE results were announced. When the beautiful graph of data appeared on the screen, the audience of staid physicists burst into spontaneous applause—a rare event indeed. The principal scientists on the project, George Smoot and John Mather, received the Nobel Prize in 2006.

From the point of view of cosmologists, COBE established an extremely important fact about the CMB. No matter which way you point the receiver, no matter how far apart the sources of your microwaves are, the temperatures associated with the black body radiation you see will be practically the same. Every part of the universe, in other words, appears to be at the same temperature. In the language of physics, we say that different parts of the universe are in thermal equilibrium. In essence, the universe is like a rub of water that has been sitting long enough for all the temperature differences between different parts to have been washed out. Deviations from this uniformity exist, but they are small—typically a small fraction of a percent. For the moment, let's concentrate on the uniformity.

The fact that the universe is in thermal equilibrium now means that at some time in the past the different parts of the universe must have been in contact with each other. And that's where we run into problems. If you run the film backward for the standard Hubble expansion, assuming the expansion to have been uniform since the Big Bang, there is no way for regions now at different ends of the universe to have been in contact long enough to establish thermal equilibrium. The only way around this difficulty, short of abandoning some well-verified laws of physics, is to question the assumption that the expansion of the universe has been uniform since the start.

This is exactly what the process of inflation does. It tells us that in the first fraction of a second of its life, the universe went through a period of rapid expansion, a process that has not been seen since then. The idea is that for the first $10^{-35}$ seconds of its life it was much smaller than we would think if we just extrapolated the present Hubble expansion backward in time. Small enough, in fact, to establish thermal equilibrium—a thermal equilibrium that was maintained during the inflationary expansion and can be seen today.

Before we move on to the formation of galaxies, we should stress that the appearance of thermal equilibrium was not the motivation for the development of the theory of inflation. That theory depends on a fairly complex series of theoretical and mathematical ideas. Thermal equilibrium plays more or less the same role in the theory of inflation that the advance of the perihelion does in general relativity. It supports the theory but was not the motivation to produce it in the first place.

## THE GALAXY PROBLEM

We have reached the point where atoms have formed and the universe has become transparent, and we are ready to move on to the next stage in the evolution of the universe. How do we proceed from here? The first step, obviously, is to ask how stars and galaxies can be formed from a hot, expanding gas. The way to approach this problem was first laid out by the British physicist Sir James Jeans (1887–1946) in the 1930s.

In any collection of matter like the interstellar clouds that are seen everywhere in the Milky Way there will be, just by chance, some regions that have more matter in them than others. Because of their increased mass, these regions will exert a greater gravitational attraction on surrounding matter than less massive areas. Consequently, they will pull in more surrounding matter, and their mass will increase. This in turn will increase the gravitational force they can exert on surrounding matter, which will increase their mass, and this in turn will further increase the gravitational force they exert.

The regions that are only slightly denser than their surroundings will continue to grow, and eventually the starting cloud will break up into dense clumps of matter that continue to contract under the influence of the gravitational attraction each part exerts on the others. We have good data to support the notion that this so-called Jeans instability is the process by which stars and planetary systems form out of interstellar clouds in the Milky Way. Our telescopes, both orbiting and ground-based, have documented this sort of stellar formation going on all around us. The question is this: Can the Jeans instability account for the formation of galaxies?

The problem is that there were two competing processes going on in the universe once atoms formed. One was the collection of material into stars and galaxies via the kind of collapse suggested by Jeans, and the other was the Hubble expansion. The former process tended to pull material together, but the latter tended to increase the separation of atoms and slow down the formation of stars and galaxies. The question is which one eventually won. The answer to this question, based on straightforward mathematical calculations, is that there simply wasn't enough time for the force of gravity to create anything like our current universe before matter is spread out too thin for massive objects to form. Thus, having traced the history of the universe from the very beginning to the point where we can imagine something like current structures to form, we are stopped by the simple inability of gravity to act strongly enough to overcome the outward tug of the Hubble expansion.

Now what?

## DARK MATTER

The resolution of this problem came from a wholly new source. In the 1970s, American astronomer Vera Rubin (1928–2016) was engaged in a study of the rotation of galaxies. Her main interest was the speed at which stars moved at various distances from the galactic centers. At the time the astronomical community had a pretty good idea of what to expect from this sort of measurement. Rubin expected to see the rotation of galaxies split into three regions. Near the galactic center, where gravitational force between stars was very strong, it was expected that those forces would basically lock the stars together, so that they all rotated together, something like a merry-go-round. This close-in rotation style was called "wheel flow."

Farther out, they expected to encounter a region where all the stars moved at the same pace. In this situation, the farther away from the center a star was, the farther it had to travel, which means that the outermost stars would fall behind those close in. A familiar example of this sort of arrangement can be seen in track meets where runners on an elliptical track have their starting positions staggered to compensate for the fact that those in the outer lanes have farther to run than those in the inner lanes.

Finally, once stars were far away from the galactic center—far enough away so that their motion could be regarded as being analogous to the orbiting of planets in the solar system—we expect to see a third type of rotation pattern. Just as Jupiter moves more slowly in its orbit than Mercury, astronomers expected to see a situation in which the farther away from the center a star was, the slower its orbital velocity would be. This expected pattern was called "Kepler flow" after the German scientist Johannes Kepler, who first deduced the orbits of planets in the solar system.

When Rubin started measuring the rotation curves of galaxies, she did indeed see wheel flow near galactic centers as well as the constant velocity flow farther out. Farther away from the galactic centers, however, she ran into a surprise. Instead of segueing into Kepler flow, the orbital velocities showed to hint of slowing down. To this day this transition to Kepler flow has not been seen.

The only way to explain this unexpected turn of events was to say that the galaxy was enclosed in a large sphere of some material that (1) did not interact with electromagnetic radiation and was therefore invisible and (2) was capable of exerting a gravitational force. Once Rubin's results were known, evidence for this new kind of matter began popping up all over the place. The velocity of stars in clusters, for example, had long been known to be too high for the stars to remain close together. One response to this situation was to say that the stellar clusters were simply temporary structures, stars caught in the act of separating. That explanation had always seemed to be a bit contrived, but this new material, with the gravitational force it exerted, could solve the problem in a natural way.

Because it was invisible, the new material came to be called "dark matter." A good way to think of it is to imagine that Earth was always enveloped in clouds, so that we could never see the Moon. We would still know the Moon was there because of its gravitational effect; the tides would betray its presence. In the same way, dark matter reveals its presence through the gravitational force it exerts, even though it can't be seen.

As surprising as it was at first, dark matter has come to be a normal part of the universe in our minds. It makes up about a quarter of all the mass/energy in the universe. Compared to ordinary matter, which makes up less than 5 percent of the universe, it plays an important role in the cosmos.

It also solves the problem of galaxy formation we discovered earlier. Because it does not interact with electromagnetic radiation, it can come together in a Jeans-type process before atoms form. It will not be affected, in other words, by high-energy photons in the primal plasma. This means that when atoms form and the photons are released, ordinary matter finds a universe in which collections of dark matter capable of exerting strong gravitational forces are already present. Instead of having to go through the laborious process of gravitational collapse, it can fall into holes already created by collections of dark matter. Thus, it has plenty of time to create stars and galaxies.

To honor Vera Rubin's achievement, Congress officially named a major astronomical observatory being built in the Andes Mountains in Chile the "Vera Rubin Observatory."

For all of the evidence for the existence of dark matter, there remains a nagging question: What, exactly, is it? As we saw in chapter 14, despite repeated attempts, no one has yet been able to produce evidence that allows us to answer this question.

Once we have this cosmological timetable laid out, we can turn to the question of the role that black holes play in the formation of our familiar universe. The answer to this question depends on the kind of black hole we're talking about. The familiar stellar black holes—the first ones we discovered—obviously can't start to form until massive stars have had a chance to run through their life cycle and collapse. This means that we don't expect to see stellar black holes until about 400,000 years after the Big Bang.

The other kind of black hole we know about—the supermassives—are a bit more mysterious. We really don't know how they form. What we can say, however, that many of our current theories about their formation require the existence of atoms, so, like their stellar brethren, they probably won't show up until massive stars have had a chance to go through their life cycle.

And what about the kinds of black holes that are theoretically possible but have never been seen, like quantum black holes? Since we've never seen these objects, it's really hard to nail down their properties. All we can really say with confidence is that they can probably show up as soon as there's enough energy in the space-time continuum to form them.

We now have a reasonably well-documented timeline into which we can start to fit we things we are learning about black holes.

# 20

## TELESCOPES IN ORBIT

---

### A man has to see, and not just look.

*Louis L'Amour,* The Quick and the Dead

I n chapter 5 we talked at length about the difficulties imposed on telescopic observations by the fact that incoming light has to travel through Earth's atmosphere. We pointed out that there is a simple solution to this problem: put the telescope into orbit above the atmosphere. In fact, by the start of the twenty-first century we have done exactly that. Sometimes, as with the Chandra X Ray Observatory, this has opened up a completely new region of the electromagnetic spectrum to scientists. In other cases, most notably with the Hubble Space Telescope, it has simply allowed us to see what's out there with greater clarity.

It's easy to say something like "put the telescope into orbit," a lot harder to do in practice. In this chapter we will walk you through the long process that led to our current best orbiting observatory, the James Webb Space Telescope. The takeaway lesson from this chapter should be that if we want to uncover the deepest secrets of the universe, we need men and women who can fight the battles in obscure committee rooms that keep massive projects alive so that scientists can have the tools they need to pry out nature's secrets.

We laid out all the disadvantages of using telescopes on the surface of the earth and having to deal with the planet's atmosphere in chapter 5. We also outlined some of the early attempts to put instruments above the atmosphere to detect electromagnetic waves that would otherwise be absorbed before they got to the surface. We can trace the notion of putting a general use telescope into orbit back to 1946 at least. In that year the American astrophysicist Lyman Spitzer (1914–97) wrote a paper titled "Astronomical Advantages of an Extraterrestrial Observatory" that caught people's attention. This was a time when astronomers were just beginning to mount instruments like X-ray detectors on rockets and see aspects of the sky that had never before been visible. In 1962, for example, NASA launched the Orbiting Solar Observatory to obtain ultraviolet, X-ray, and gamma ray data from the Sun. Little by little, the idea of putting a major telescope into orbit started to catch on in the astronomical community.

Two scientists were especially prominent in campaigning for an orbiting space telescope in those early days. One was Spitzer, and the other was Nancy Grace Roman (1925–2018), the chief of astronomy at NASA in the 1960s and 1970s. Roman is often referred to as the "Mother of Hubble." Both have been honored by having orbiting observatories named in their honor; the Spitzer telescope was an infrared instrument that was operated from 2003 to 2020, and the Roman telescope is an infrared survey instrument scheduled to be launched in 2026.

By 1977 the idea of building what was then called the Large Space Telescope (LST) had gained wide acceptance in the astronomical community. Getting funding for the project from Congress turned out to involve some unusual political shenanigans in which NASA dodged a bullet and kept the Hubble Space Telescope alive. In the original budget proposal NASA asked $5 million for the space telescope project, more or less as a kind of placeholder; it wasn't enough to start serious work and was largely viewed as an attempt to pacify the astronomical community. Instead, other NASA administrators decided on a different strategy. They took the $5 million out of the budget proposal, essentially telling the astronomical community that unless they started some serious lobbying they would lose the LST completely. Here's how NASA Associate Administrator for Space Science Noel Hinners described the strategy:

> It was clear that year that we weren't going to be able to get a full-up start. There was some opposition on [Capitol] Hill to getting a new start on [Hubble]. It was driven, in large part as I recall, by the budget situation. Jim Fletcher (the NASA Administrator) proposed that we put in $5 million as a placeholder. I didn't like that idea. It was, in today's vernacular, a "sop" to the astronomy community. "There's something in there, so all is well."
>
> I figured in my own little head that to get that community energized we'd be better off zeroing it out. Then they would say, "Whoa, we're in deep trouble," and it would marshal the troops. So I advocated that we not put anything in. I don't remember any of the detailed discussions or whether there were any, but Jim went along with that so we zeroed it out. It had, from my perspective, the desired impact of stimulating the astronomy community to renew their efforts on the lobbying front. While I like to think in hindsight it was a brilliant political move, I'm not sure I thought it through all that well. It was something that was spur of the moment. . . .
>
> $5 million would let them think that all is well anyway, but it's not. So let's give them a message. My own thinking, get them stimulated to get into action. Zeroing it out would certainly give

that message. I think it was as simple as that. Didn't talk to anybody else about doing it first, just, "Let's go do that." Voila, it worked. Don't know whether I'd do that again.

It was unorthodox, but it worked. There has always been a whole lot more to advancing science than just working out technical details, and this kind of political gameplaying is always going to be needed. Fortunately, from the time of the World War II Manhattan Project to the present, there have always been a few politically savvy scientists ready to go into battle in situations like this.

The Large Space Telescope was renamed in 1977 to honor the great American astronomer Edwin Hubble. Unlike the later experience with the Webb, there was very little controversy about the name of the first space telescope. Work proceeded apace, and launch was schedule for 1983. The usual kinds of delays moved the launch date to 1986.

From the beginning it was planned to put the Hubble into low Earth orbit, 340 miles above the ground. This turned out to be a real blessing, since it allowed astronauts to visit the telescope to do maintenance work and upgrade the detection systems. Unfortunately, before the telescope could be loaded on to the space shuttle and launched, the shuttle *Challenger* exploded during its takeoff and the entire shuttle program was shut down for two years. During this waiting period the telescope had to be kept powered up in a clean room, at the cost of some $6 million per month. Nonetheless, the delay allowed engineers to improve some of the telescope's systems before launch, which finally happened in 1990. The cumulative cost of the Hubble to date, including servicing missions, was over $11 billion, which makes it one of the most expensive science projects ever.

But getting into orbit did not solve all of the Hubble's problems. When astronomers began pointing the telescope at the stars, they saw a faint halo around their targets. Investigation revealed that there was a serious problem with the main mirror: it had been ground precisely, but to the wrong shape. It wasn't a huge error—the outer edges on the main mirror had been ground down one eleven-thousandth of an inch too much, but that was enough to throw the instrument off. The culprit was eventually identified as the incorrect assembly of one of the optical instruments used to check the shape of the main mirror as it was being ground. (It wasn't, as some wags suggested at the time, that the mirror makers had used English as opposed to metric units in the assembly process.)

What to do?

There were more options that you might think. There was, for example, a spare mirror that had been built as a "just in case" replacement for the Hubble. There was the option of bringing the telescope back to Earth and fixing the problem in a terrestrial laboratory. Of course, in that case there was no guarantee that Congress would ever let it fly again. Consequently, NASA undertook what had aways been plan A: sending a team of astronauts up to replace some of the Hubble's optics. The patch job worked.

Listing all of the accomplishments of the Hubble telescope would take more room than we can spare in a book like this, so we'll talk about just one image, the so-called Hubble Deep Field. This image was produced by looking at a patch of the sky the size of to the head of a pin held at arm's length, a patch with very few stars and galaxies visible to ordinary telescopes. Hubble was trained on this remote bit of the sky for a total of more than eleven days, and in the end it produced images showing more than three thousand faint and therefore distant galaxies. Suddenly our picture of the early universe and the formation of galaxies changed overnight. In this book, for example, we have thrown out phrases like "billions of galaxies" without bothering to explain where that number came from: it's just part of our current astronomical received wisdom. Now we understand that it comes from extrapolating the result of the Hubble Deep Field over the entire sky, a fairly straightforward exercise.

As we write these words in the spring of 2024, the Hubble is still in operation, although it is showing signs of age. Problems are developing with the gyroscopes that are used to point the telescope axis, and NASA has no plans to send another team of astronauts to refurbish the system. We expect that Hubble will continue to do forefront science right up to the end.

## THE JAMES WEBB SPACE TELESCOPE

Discussions were going on in the astronomical community to build the successor to Hubble before the Hubble was even launched. This shouldn't come as a surprise, given the incredibly long time it takes to go from a good idea to an actual piece of hardware. The reasons for the discussions, however, didn't have so much to do with politics and finance as they did with important changes that were taking place in the scientific community. People knew about the Big Bang in the mid-twentieth century, but the study of cosmology and the early universe was not very well advanced at that time. Cosmologists were just beginning to assemble data that allowed them to provide quantitative answers to questions such as those involving the evolution of stars and galaxies.

The only way to get answers to these kinds of questions is to observe the stars and galaxies that formed soon after atoms formed and the cosmic background radiation was released (see chapter 19). We need data on the most distant stars and galaxies that we can possibly detect, in other words.

This creates a series of problems if you are designing a telescope. For starters, the objects you are looking for will be very faint, which means that you will need a telescope with a mirror much larger than the one on the Hubble. Since the limit on the size of that mirror was set by the size of the space shuttle cargo bay, we can't solve this problem by just increasing the size of the mirror. A new technology is needed.

There is a more fundamental problem created by the nature of the Hubble expansion itself. The light from these objects has been traveling toward us for roughly 13 billion years, and during that transit the stretching of space by the Hubble expansion has shifted the wavelength of the incoming radiation. What started out as visible light at the newborn galaxy will have been shifted into the infrared by the time it gets to us. This means that, unlike the Hubble, which sees primarily in the range of visible light, its replacement will have to see in the infrared.

As we pointed out in chapter 7, infrared telescopes face a fundamental problem created by the laws of physics. The fact is that every object in the universe at a temperature above absolute zero (which means, in essence, every object in the universe) emits some sort of electromagnetic radiation. These emitters include the telescope, which can, in effect, detect its own radiation and "see itself." The only way around this problem is to lower the temperature of the telescope to the point that the radiation it is emitting can be distinguished from the radiation emitted by an object being observed. The first orbiting infrared telescopes dealt with this problem by carrying a cooling liquid (typically liquid helium) to maintain a low temperature. This solution worked, but it had the disadvantage of limiting the useful life of the observatory, which would stop functioning when the coolant ran out. Clearly, if you are spending billions of dollars to build a telescope, you don't want to limit its useful lifetime this way.

In 1994 a group called "HST and Beyond" committee was formed. In its 1996 report to NASA, the committee discussed these problems and proposed the development of a cold infrared sensitive orbiting observatory. This began the planning process for what eventually became the James Webb Space Telescope. The next major step was triggered by what is called the 2000 Astronomy and Astrophysics Decadal Survey. It has become the custom for various fields of science to present a list of what its practitioners see as the major needs for advancing that particular field. This list is typically presented as a prioritized list to Congress and various funding agencies. These lists, which are updated every ten years (hence "decadal") are intended to serve as a guide for funding decisions. Compiling the list allows scientists to say to Congress and various foundations "here are the facilities we need to advance in this field." In 2000, what was then called the Next Generation Space Telescope topped the list. It was seen as a way to study the formation of stars and galaxies in the early universe as well as make detailed studies of exoplanets as part of the search for extraterrestrial life. The Next Generation Space Telescope topped the list of that particular survey and was seen as the most important instrument needed for the advance of astronomy. The proposal was accepted overall, and eventually the telescope was named after James Webb, a former NASA administrator.

The launch of the Webb began to seem like a real possibility. One issue was where to park the telescope once it had been launched. Over the years space scientists have learned to use special points in space to place satellites. They are called Lagrange points after the mathematician Joseph-Louis Lagrange (1736–1813), although the mathematician Leonhard Euler (1707–83) had discovered some of them earlier. Lagrange points are places in the region of two gravitating bodies where the gravitational and centrifugal forces balance and cancel each other out. There are five Lagrange points in the Earth/Sun system. Three of these points are located on a line between the centers of Earth and the Sun (these are the points that Euler discovered) and are labeled $L_1$, $L_2$, and $L_3$. The other two, off to the side, are each located at the apex of an equilateral triangle with the Earth and Sun being the other two vertices of the triangle.

The most important Lagrange points are $L_1$, located between Earth and the Sun and $L_2$, located on the line between centers but outside the orbit of Earth. $L_2$ is particularly good spot to place satellites and space observatories. Since the second Lagrange point is far outside the orbit of the moon, the solar panels of all of those systems are never in shadow but are constantly in direct sunlight. Furthermore, although the Lagrange point is not actually stable, it is close enough to stability to require very little fuel to maintain a satellite's position.

There is another important advantage to parking the James Webb Space Telescope at the second Lagrange point. This Lagrange point is far enough from the Sun, the Moon, and Earth that all radiation from those bodies can be blocked by a single screen. Thus, the telescope can be shielded from heat generated by all three bodies.

These advantages have not been lost on the astronomical community. Several important satellites besides the James Webb have been parked at $L_2$. The European Space Agency has its Gaia (dedicating to measuring the position of stars with unprecedented accuracy) and Euclid (dedicated to measuring the expansion of the universe) satellites in orbit around the second Lagrange point. $L_2$, together with its entourage of satellites, follows Earth as the entire assembly orbits the Sun. The net effect is that the James Webb telescope is always pointed outward, available to take data, while it is maintained at the temperature of deep space by the sunshield that we will describe in a moment. Another advantage is that putting the telescope at the second Lagrange point allows for direct communication and data transfer to Earth.

One disadvantage of this arrangement is that $L_2$ is a bit less than a million miles from Earth, far too distant to accommodate repair and refurbishing missions crewed by astronauts. In addition, NASA engineers have concluded that robotic missions to $L_2$ are beyond the capability of current technology. In other words, once the James Webb is parked at the second Lagrange point, that's it: no repairs, no changes, no fixes.

We thus have to build a mirror bigger than the one in Hubble and fit it inside the restricted space of the cone of a rocket. We have to design a sunshield to maintain the telescope temperature and fit *that* into the same rocket ship. We have to fly all that stuff to a point a million miles away and unfold it flawlessly before we get a single picture.

Did we mention engineering nightmare?

Let's start with the mirror, the main "light bucket." Here technical developments on ground-based telescopes gave the Webb designers a head start. In chapter 7 we discussed the advantages of segmented mirrors controlled by active optical systems. This technology allows ground-based telescopes to have much larger mirrors than would be available from a mirror cast from a single piece of glass. Using this technology allowed the Webb design team to create a mirror six times larger than that on the Hubble, but to make it from small enough segments so that the mirror could be fit into a rocket ship. In fact, the Webb mirror is made from eighteen hexagonal segments, each about 1.3 meters (a bit more than four feet) across. These segments can be folded together during launch and unfolded when the rocket reaches its destination. This arrangement gives the Webb a primary mirror 6.5 meters (21 ft.) in diameter.

That may solve the problem we started with, but it creates a new one: How can we be sure that the mirror can be folded up for launch and then unfolded when the Webb has reached $L_2$? NASA engineers dealing with this problem consulted Japanese experts in the art form of origami, the art of folding paper. Those engineers designed a system in which there were two "wings" to the arrangement of hexagonal mirrors, each wing with three mirrors stacked vertically. Before launch, the two side wings were folded over so that the main mirror was narrow enough to fit into the rocket. They were then folded back out when the telescope was in its final position. The Webb main mirrors are made of beryllium, a light, strong metal capable of withstanding large temperature changes, covered by a thin layer of gold to increase reflectivity and a thin glass sheet for protection.

If you think the deployment of the mirror sounds complicated, the sunshield is more complicated still. The basic strategy of keeping the telescope temperature down depends on the deployment of a shield to keep heat emitted by the Sun, Earth, and the Moon from reaching the telescope. Given the variations in position of those objects and the telescope, the sunshield in front of the telescope would have to be about the size of a tennis court. Obviously, this is going to cause another folding problem.

The sunshield has a huge job to do. On the side facing the Earth/Sun/Moon system, the temperature will be around 185°F (85°C), but on the shaded side it will be −188°F (−233°C). Obviously, dealing with this sort of temperature difference is a difficult job.

The sunshield consists of five separate layers of metal coated plastic, each sheet the thickness roughly that of a human hair. Aluminum coating increases reflection, and the Sun-facing layer has a coating of doped silicon to do so as well. This has the effect of making the color of the shield a most unscientific-looking pinkish-lavender. Each layer of the shield deals with the radiation that the outer layers didn't pick up. The vacuum between the sheets prevents heat from being conducted into the telescope itself. The sheets are arranged so that they are separated by a few inches at the center and about ten inches at the edge. This has the effect of moving heat out into space from the edges of the shield.

Deploying the shield is a very delicate job: the thin sheets need to be furled out one by one, and during a deployment test in 2018, one of the sheets tore. This was one of many causes for the delay in the launching of the James Webb. The sunshield will be subject to damage from meteors in space. This can't be avoided, but the plastic sheets are covered with reinforcement every six feet so that any meteorite damage is contained in a small area.

The plastic used in the sheets is called Kapton. It was developed by Dupont Corporation in the 1960s and is widely used in electronic systems. This is one more example of how terrestrial developments have played an important role in the construction of space telescopes.

Imagine that it's 7:00 in the morning of December 25, 2021, and you're at the European Space Port in French Guiana. An Ariane 5 rocket, as dependable a rocket as there is, is on the launch pad, and inside it, folded up, is the $10 billion James Webb Space Telescope. Your engineers have told you that there are no fewer than 344 different ways this mission could fail. The rocket lifts off at 7:20, and the telescope begins its twenty-nine-day journey to its home at the second Lagrange point. The computer screens show the rocket leaving orbit and heading out, the last time any human being will see it.

What would you have given to have seen that launch?

## ORIGINS

Ever since 1924, when Edwin Hubble showed that the matter in the universe was ordered into galaxies, those galaxies have been a problem. Actually, the "galaxy problem" is a sequence of several problems, so that as soon as one is solved another pops up. Let's look at them in sequence.

*Galaxy problem 1: Breaking the homogeneity.* The central issue of the first iteration is easy to state: we believe that the universe started out as a homogeneous mixture, but it is not homogeneous now. In chapter 19 we saw that the earliest stages of the Big Bang produced an expanding universe composed of a roiling sea of particles and radiation. The universe was turbulent, but it was homogeneous. By this we mean that one chunk of the universe was pretty much like every other chunk. There were no distinct structures anywhere and, as we have seen, if any structure—even a simple one like a complex atomic nucleus— tried to form it would quickly be blown apart by the radiation.

That universe was quite different from the one we live in today. When we look out at the sky, we see almost nothing but structure. Planets and stars dominate our attention, but if we had to name a single special building block of our universe, we would surely pick galaxies. And that's where the problem begins, because there are many different kinds of galaxies, and they seem to pop up in large structures that span an appreciable fraction of the sky.

Understanding this, we can redefine the first version of the galaxy problem as follows: How did the universe transition from a homogeneous ball of matter and radiation to the highly structured system we see today? And how did it make that transition in the relatively short time between the Big Bang and the oldest stars and galaxies our instruments can detect?

The first theorists to tackle this problem in the mid-twentieth century started by assuming that something like the simple process of gravitational collapse we discussed in chapter 4 would solve the problem. This is a process in which matter in an interstellar cloud is pulled into regions of higher density, eventually reducing the cloud to a series of dense objects like stars. This starting assumption seemed like a good idea. After all, we know that gravity acts in the same way throughout the universe, so why not let it form galaxies as well as stars?

Unfortunately, this way of approaching the birth of galaxies neglects the other important process operating in the early universe: the Hubble expansion. While gravity is carrying out its leisurely collection of matter into galaxy sized clumps, the expansion is carrying matter away from the places where gravity is trying to pull it together. If you do the numbers, you find that if gravity is the only force available to you, it simply acts too slowly to create the universe we see. This, in essence, was the second iteration of the galaxy problem.

*Galaxy problem 2: Dark matter.* This is why the discovery of dark matter (see chapter 19) was so important. The one thing we know about dark matter is that it interacts with gravity, but not with electromagnetic radiation. This means that dark matter can start clumping together much earlier than ordinary matter can. Consequently, when atoms form and the universe becomes transparent, ordinary matter finds itself in a situation in which clumps of dark matter are already exerting strong gravitational forces. In essence, all ordinary matter has to do to form galaxies is to fall into the gravitational holes created in advance by dark matter.

Jim Trefil ran into an amusing incident associated with the discovery of dark matter. He wrote a book called *The Dark Side of the Universe* in which the first chapter, which laid out the galaxy problem, was titled "Why Galaxies Can't Exist." In a public debate with a creationist, he found his opponent quoting that chapter as proof of the failure of conventional science. Sadly, Jim's pointing out to his interlocutor that the other fourteen chapters in the book showed how the discovery of dark matter solved all the problems posed in chapter 1 failed to register. Ignoring data is, unfortunately, a common debating tactic for creationists.

With the discovery of dark matter, we were able to put together a theoretical framework that does a pretty good job of dealing with galaxies and the large-scale structures of which they are part. It's called the $\Lambda$CDM model. ($\Lambda$ is the Greek capital lambda, corresponding to $L$ in the Latin alphabet.) $\Lambda$ is called the cosmological constant, customarily used to refer to quantities associated with dark energy, such as the cost of creating more space (see chapter 19). CDM refers to "cold dark matter," where "cold" refers to the fact that the dark matter—whatever it is—is not moving at relativistic speeds.

The formation of single galaxies is not the only galaxy "problem." Once galaxies form they create a very complex structure, a structure that is far from homogeneous. The best way to visualize what we see out there is to picture a big pile of soap suds and then to imagine cutting through it with a knife. What you'll see is that the soap is located on the walls of a complex set of bubbles. In the same way, galaxy clusters are found at the edges of vast empty spaces called voids. The central question is: How did we get from a collection of single galaxies to a universe filled with structure?

The way that theorists investigate a question like this is to create computer models that show the kinds of galactic patterns that will result from making specific assumptions about things like the behavior of dark matter and then compare those computer-generated patterns to what we actually see. When this is done with the $\Lambda$CDM models, the results loo pretty good, so for a while, at least, cosmologists could relax and think that they had pretty well solved the galaxy problem.

Then came the James Webb Space Telescope.

*Galaxy problem 3: The James Webb Space Telescope results.* It is a truism that looking far out into intergalactic space is also looking back in time. That is one reason why instruments like the Hubble and James Webb space telescopes play such an important role in our quest to understand the universe. It has been 13.8 billion years since the Big Bang, and it is not at all unusual for our space-based telescopes to probe the universe as it was 13 billion years ago, less than a billion years after the universe was born.

One thing we know from our study of nearby galaxies is that there is a strong connection between galaxies and supermassive black holes. We know, for example, that virtually all galaxies have supermassive black holes in their nuclei, and that the bigger the galaxy, the bigger the black hole. It makes sense, therefore, to focus our attention on supermassive black holes as we set out to resolve the third iteration of the galaxy problem.

To go as back in time as they can, astronomers often use a trick known as gravitational lensing. Here's how it works: suppose the object you want to examine is a bit too far to be imaged directly by your telescope, but suppose that there is a massive object in the line of sight to your target. We know that massive objects can bend the path of light rays—indeed, detecting this deflection was what brought general relativity to the fore in popular and scientific consciousness. This means that light rays from your target object will be bent around as they lasso the intervening mass. In effect, that mass focuses and concentrates the light rays coming from the target object. The intervening mass, in other words, acts like a lens, focusing the light just as the glass lens in a microscope does. The net effect is that gravitational lensing extends our ability to see and analyze distant objects.

There are several points we can make when we talk about examining the universe as it was in the first billion years of its existence. For one thing, electromagnetic radiation of all types will be shifted—visible light, for example, will be shifted into the infrared by the Hubble expansion. This is why the James Webb Space Telescope (JWST) was designed to detect infrared radiation rather than visible light. For another thing, the galaxies we examine will be so far away that there will be no hope of seeing the details that we have used up to this point to show the existence of supermassive black holes in galactic nuclei. Instead, we will have to detect radiation (typically X-rays) given off by the accretion ring of a suspected supermassive. The searches are done with a team of telescopes: the JWST to see the infrared light, and the Chandra X-ray observatory to see the X-rays. The ability to carry out this kind of tandem search illustrates the wisdom of maintaining a fleet of orbiting observatories above Earth's atmosphere.

In 2023 NASA announced the discovery of the oldest known galaxy with a supermassive black hole at its center—a system that is estimated to have been born about 470 million years after the Big Bang. The supermassive is in a galaxy called UHZ1, located a bit more than 13 billion light-years from Earth. A galactic cluster called Abell 2744 is located on the line of sight between Earth and UHZ1, some 3.5 billion light-years from Earth. This cluster acts as the gravitational lens that allows us to collect enough light to see the much more distant supermassive. (The gravitational lens supplied by Abell 2744 magnifies the image of UHZ1 four times.)

Once the JWST located the galaxy containing the supermassive, the Chandra observatory spent two full weeks putting together an X-ray picture of the accretion disk of the black hole. The fact that this much time on a major observatory was allotted to this measurement is a clear signal that finding structures close to the Big Bang is a clear indication of how important astronomers think this data is.

The supermassive black hole in UHZ1 is estimated to have a mass somewhere between 10 and 100 million times the mass of the Sun. It appears to have a mass equal to that of all of the stars in its home galaxy. (Another way of saying this is that half of the mass of this particular galaxy is locked up in its central black hole.) It is just one of many supermassive black holes that seem to have existed in the first billion years of the life of the universe.

Perhaps the most interesting detection story associated with the early black holes concerns a system labeled GNz7q, which was present some 750 million years after the Big Bang. This system was not discovered by a dedicated search, like UHZ1, but had been hiding in plain sight in photographs taken by the Hubble Space Telescope years ago.

The discovery of all of these early black holes raises what we can call the ultimate form of the galaxy problem. How did so many supermassive black holes show up so soon?

There are basically two ways to approach this question. One is to note that there is a well-understood sequence of events that could lead to a supermassive black hole. Large stars forming early in the life of the universe could go through the standard main sequence/supernova/ black hole cycle, producing a collection of stellar black holes. These would then go through a series of collisions like the one described in chapter 19, eventually producing a supermassive black hole. This approach has the advantage of familiarity, but it's hard to see how it could produce a supermassive like UHZ1 in a time frame consistent with the observed lifetime of that black hole.

The other option is to assume that somehow the black holes form by some sort of direct collapse process from the material available in the early universe. Some variations of this sort of suggest that largish seeds are formed early on, and then the growing black hole simply accumulates matter from its surroundings. This sort of process could deal with the time constraints that the new data places on black hole formation.

It is likely that as more data accumulates about black holes that existed near the origin of the universe, this problem will get sorted out. When that happens, we will have a clear picture of the way the universe evolved from a homogenous ball of matter and radiation to the complex structure we see around us today.

# 21

# THE END OF THE UNIVERSE

———

From too much love of living,
From hope and fear set free,
We thank with brief thanksgiving
Whatever gods may be
That no life lives for ever;
That dead men rise up never;
That even the weariest river
Winds somewhere safe to sea.

*Algernon Charles Swinburne,* "The Garden of Proserpine"

Everything has to end sometime, even the universe. It is fitting that when we think about this mysterious subject we run into the most mysterious object we know about: the supermassive black hole. Getting from a universe full of blazing stars and complex chemistry to a dying collection of black holes is not a trivial task, however, and, like many topics we've tackled in this book, it's going to involve material from many different branches of science.

There are actually two different ways of approaching the end of the universe—think of them as "top-down" and "bottom-up." The top-down approach involves looking at some of the general properties of the universe, asking how they got to be the way they are, and then trying to understand how they will play out in the future. The bottom-up approach involves looking at specific points of view, such as observing the dying universe from the surface of Earth and tracing the last events from there.

Let's look at the top-down approach first. When we look at the universe we see two central processes: (1) the universe is expanding, and (2) the expansion rate of the universe is increasing. In chapter 19 we discussed these two characteristics with an automotive analogy. In this analogy the force of gravity between galaxies acted as a sort of a brake on expansion, while the force associated with dark energy acted as a kind of gas pedal. Early on, when matter was close together and gravity was at its strongest, the brakes were on, and the Hubble expansion was slowing down. Around the time that the universe was about 5 billion years old, the distances between galaxies got big enough for the gas pedal to win the tug of war, and the expansion started to accelerate. The expansion accelerated, and that's the kind of universe we're living in right now.

It should be obvious that what happens in our future depends on the interplay between gravity and acceleration. To put this into modern language, it means that what happens in our future depends on the future behavior of dark energy—the cosmic gas pedal. Here we run into a problem, because nobody has the faintest idea of what dark energy actually is. The best we can do in this situation is make a series of assumptions about the properties of dark matter and hope that somewhere in the array of possible futures we predict is the one that will actually happen.

Given that, we can see several possible futures for dark matter. The most dramatic ending would result from a situation in which the acceleration of the universe caused by dark energy increases, leading to an ever-increasing density of dark energy. Another possible future is one in which the amount of dark energy increases in proportion to the increase in the amount of space-time itself. Finally, we can consider a future in which the density of dark energy decreases as the universe expands, so that gravity eventually wins the tug of war. Each of these possibilities—and we have to stress that our current understanding of dark energy doesn't allow us to choose between them—leads to a different kind of end for our universe.

## THE BIG RIP

Let's start with a future in which the currently observed acceleration of the Hubble expansion increases. Note that this is an acceleration of the observed acceleration—quite a rapid increase. (It involves a nonzero third derivative of intergalactic distances.) The key point, as far as general relativity is concerned, is that in this situation the density of dark energy (which supplies the "oomph" to expand space-time itself) increases. As time goes by, there is more and more of the stuff out there.

Once we get into this scenario, the language we use begins to get a little weird. For example, the excess dark energy that leads to the acceleration of the cosmic acceleration is called "phantom energy." As far as we can tell, this name was taken from the Star Wars movie installment *The Phantom Menace*, which came out a few years before the scholarly papers suggesting the scenario we're about to describe.

If space-time is being pulled apart so that adjacent points are separating faster than the speed of light, then the ability of two observers (in neighboring galaxies, for example) to see and communicate with each other lessens over time. Picture a shrinking sphere of points with which you would be able to communicate in that situation. The distance to the farthest point with which you could theoretically communicate in the future is called your "cosmic event horizon." One way of thinking about the universe in which the acceleration is accelerating is to imagine your cosmic event horizon as a shrinking sphere centered on your location.

What would the universe look like in this situation? If space-time is continually pulling points apart as a speed faster than that of light (and remember, space-time itself is not bound by the "nothing can travel faster than light" rule), then the most loosely bound structures will be torn about first, but eventually everything—even atoms—will be torn apart as the space-time on which they sit expands. This scenario is called the "Big Rip."

The authors of the original paper on this subject worked out a reasonable numerical example to produce a timetable for the Big Rip. In their scheme, the Big Rip would happen 22 billion years from now. About 200 million years before the final event, galaxies would start to be separated from each other, with no more clusters and galactic structures. At 60 million years the galaxies themselves would start to come apart as the force of exerted by the dark energy overcame the force of gravity holding stars together. Three months before the Big Rip, planetary systems like ours would come apart, and the planets would go off on their own. In the last few minutes the stars and planets would themselves be torn apart, and at $10^{-19}$ seconds the resulting atoms would be torn apart. Presumably at some subsequent point the nuclei of atoms would be torn apart, and the elementary particles would revert to their nature as composite systems made of quarks. The resulting universe would be a vast desert of expanding space-time carrying along a few lonely particles forever looking for a partner with which to interact, but never finding one.

Depending on the assumptions you make about the nature of dark energy, black holes could exert enough gravitational force to overcome the dark energy, in which case they would be isolated islands of "normalcy" in that future void. Alternatively, they too could be torn apart, leaving behind only their singularities. (Such an isolated entity is called a "naked singularity.")

Any way you look at it, the Big Rip is a pretty unpleasant future. The most cheerful thing we can tell you is that most cosmologists consider it a pretty unlikely scenario.

## BUSINESS AS USUAL

There is a general consensus among astrophysicists that whatever dark energy is, it is somehow connected to the cost of creating space-time. Since the Hubble expansion—accelerated or otherwise—creates new space-time, this assumption means that we expect more dark energy to be created in the future. The extreme outcome of this expectation is the Big Rip scenario we've just explored. There are less dramatic assumptions we could make. For example, we could assume that while the amount of dark energy in the universe indeed increases in the future, the density of dark energy doesn't change. In other words, whenever a cubic kilometer is added to the universe, a proportional amount of dark energy is created as well to balance it. In this scenario the density of dark energy (which is equal to the total amount of dark energy in the universe divided by the volume of the universe) stays the same. Because this seems to be the way things are right now, we have called this the "business as usual" scenario.

The primary question we have to answer after making this assumption concerns the future of the Hubble expansion. We can imagine three options for this future. The expansion could go on forever (in which case we have what is called an open universe), it could reverse itself (in which case we would have a closed universe), or it could be at the transition point between these two possibilities—a universe in which the expansion slows down and stops after an infinite amount of time (which would be called a flat universe).

Most cosmological theories predict a flat universe, one whose expansion will glide to a stop at infinity. Nevertheless, given the choice of open, closed, and flat, it's obvious that it is the force of gravity that will determine which way the universe will evolve, and the force of gravity in turn will depend on the amount of matter in the universe. Consequently, in the late twentieth century a lot of astronomers' efforts were devoted to determining the total amount of mass in the universe. So strong was the theoretical bias toward a flat universe that this effort was dubbed a search for the "missing mass," by which people meant the mass that would have to be out there to produce a strong enough gravitational attraction to make the universe flat. Given that these searches were started before the discovery of either dark matter or dark energy, it's not surprising that they never found enough mass to do the job.

It was only after both of these unexpected sources of energy were found that our current picture of the universe, in which less than 5 percent of the total mass of the universe is made up of stuff like us, that a true accounting of the "missing mass" can be made. Our best accounting at this time is that 4.95 percent of the matter of the universe is what is called baryonic matter—material like us—while 26.8 percent is dark matter and 68.3 percent is dark energy.

The fate of a flat universe in a business-as-usual scenario is easy to predict. The universe will continue to expand and cool, as it has for the last 14 billion years. Eventually it will run into a situation in which it is no longer possible to extract useful work from the low-temperature universe. This is called the "heat death" of the universe, the subject of all kinds of gloomy writing for decades. It is nonetheless a possible end for the universe.

As the universe continues its asymptotic approach to infinity, in about 100 million years we won't be able to see neighboring galaxies, and in 100 trillion years all the available hydrogen that could serve as fuel for stars will be used up. Calculations suggest that $10^{20}$ years from now (that's a hundred quintillion), the supermassive black holes now residing in the hearts of galaxies will have sucked up the remnants of every galaxy. By $10^{100}$ years even they will have disappeared, giving up their energy through Hawking radiation.

It will be a long, dark, extremely boring end.

## THE BIG CRUNCH

Probably the least likely end for the universe, but in many ways the most interesting, is an ending in which the Hubble expansion slows down, stops, and is reversed. In some versions of the scenario all the matter in the universe comes together in a single point called the Big Crunch, at which point a sudden expansion begins a new Big Bang. This leads to a picture of an eternal, cycling universe, very different from other theories.

We say that this is the least likely ending for the universe because, as we have stressed repeatedly, our current data tells us that the Hubble expansion is actually accelerating rather than slowing down. We know that gravity will weaken as galaxies move farther apart, so in order to reverse the expansion dark energy would have to have some really strange properties. Since we have no idea what dark energy actually is, there is no real basis for rejecting any claims, however bizarre they may seem. After all, the acceleration of the Hubble expansion and the introduction of dark energy seemed pretty bizarre a short time ago.

Given that uncertainty, it is something of a relief to find that we can speak with some certainty about some of the astronomical events that will occur on the way to the Big Crunch. The first step will most likely be multiple collisions of galaxies. It may seem that the collision between two objects, each containing a hundred billion stars, is going to set off some pretty spectacular fireworks. Galactic collisions happen pretty often in our normal universe. The key to understanding them is to remember that galaxies are mostly empty space. Run one galaxy through another, and it would not be unusual to see no stellar collisions at all. The main interactions at this stage of the Big Crunch will be gravitational. Our current theories of galaxy evolution suggest that most of the elliptical galaxies we see are the result of precisely this kind of collision.

Eventually, the density of matter will become so great that stars will have no choice but to collide with each other. Presumably this will happen long after the stars have used up the available fuel for their nuclear fires, so the likeliest result will be the creation of a lot of stellar corpses destined to become the fuel for supermassive black holes—supermassives much larger than any we see out there today. As these galaxy-sized supermassives go into their death spiral, you can imagine the resulting black hole swallowing everything left outside of its Schwarzschild radius.

After that we are at the mercy of the astrophysical theorists, who have devised numerous ways that things might go. Here's is a sampling of the kinds of theoretical descriptions of Big Crunch cosmologies:

- Physicist Paul Davies of Arizona State University worked out a timetable for the Big Crunch, estimating when each size star would be dismantled and become part of the growing accumulation of hot material . For example, his theory suggests that the largest stars would be taken apart approximately 100,000 years before the Big Crunch.
- Physicist Paul Steinhardt of Princeton University has published a complex scenario in which the universe is dominated by two multidimensional sheets known as "branes." In his scheme, our four-dimensional universe is one brane, and both the Big Crunch and successive Big Bangs result from collisions between branes in higher dimensions.
- Roger Penrose, who shared the 2020 Nobel Prize in physics for his work on black hole singularities, has proposed a theory that goes by the name of conformal cyclic cosmology. In this scheme the cosmic expansion leads to a situation where all matter created in a Big Bang

eventually becomes massless radiation, leading to the beginning of a new cycle.

Theoretical descriptions of Big Crunch scenarios tend to be highly abstract and mathematical, even though it is not at all clear that they will ever actually occur.

## THE VIEW FROM EARTH

One way of appreciating the role of supermassive black holes in the end of the universe is to imagine standing on Earth, suitably protected and immortal, and watching the scenario unfold. We'll start our story in the present, with Earth pretty much as we see it every day.

As we saw in chapter 4, the Sun is in what is called its main sequence phase, burning hydrogen in its core to generate the outward pressure needed to balance the inward force of gravity. The Sun has been burning hydrogen for about 4.5 billion years and has enough hydrogen in its core to last another 5 billion years or so.

This doesn't mean that we can count on 5 billion more years of steady sunshine, however. The Sun is actually getting hotter as it ages. It is, for example, about 30 percent brighter today than it was when it was born, and this slow heating will go on throughout the time the Sun stays on the main sequence—another 5 billion years, in other words. We don't expect the surface of Earth to change much in the near future. The motion of tectonic plates on Earth's surface won't be affected by the temperature change, and scientists expect that about 250 million years from now all the continents will be joined together in one massive land mass. This has happened before, and the resulting supercontinent is customarily referred to as Pangea (meaning "all Earth" in Greek).

In about a billion years or so, the mean temperature of the planet will have risen above 100°F, and strange things will start to happen. The increased temperature will cause the evaporation of water from the oceans. The water molecules in the atmosphere will be broken up into hydrogen and oxygen molecules by ultraviolet radiation from the Sun, and the lightweight hydrogen molecules will wander off into space. The oceans will eventually disappear, and the planet will turn into a hot, dry world totally unlike the one we're living in today. Volcanoes will continue to pump gases into the atmosphere, but without an ocean to absorb them the surface temperature will shoot up as Earth enters a runaway greenhouse phase. Eventually, a few billion years into the future, the surface temperature will rise above the melting point of surface rocks on the planet, and the entire surface will become a lava ocean.

The final act for the planet will actually depend on the evolution of the Sun. Five and a half billion years from now, when the hydrogen in the core has been depleted, there will be two sources energy left for our star. One source is a shell of unburned hydrogen just outside the core, and the other is the helium that the Sun produced by burning the hydrogen in its core. As the fusion-driven pressure starts to drop, the relentless force of gravity reasserts itself, and the interior temperature rises as the Sun contracts. Eventually it gets to the point where the unburned hydrogen will ignite, after which the helium will start to fuse: three helium nuclei will come together to form a carbon nucleus, generating energy in the process. The Sun would be in the process of becoming a red giant star.

From the point of view of our observer on Earth, two important things happen during this transition. For one thing, the solar wind becomes very strong, and the Sun loses about a third of its mass. This lowers the gravitational force between Earth and the Sun, and as a consequence the planet's orbit moves further out. At the same time, the Sun's outer layers swell enormously, eventually reaching the present orbit of Earth. Whether Earth is swallowed up or becomes a burnt-out cinder orbiting the Sun depends on the details of Earth's orbit, but the planet will no longer be capable of supporting life. In the words of astrophysicist Neil deGrasse Tyson, this would be a good time to ask how the space program is going.

The next event that will capture the attention of our terrestrial observer will happen about 5 billion years from now. The Andromeda galaxy will start to transform from a distant fuzzy patch of light in the sky into a recognizable collection of stars. We will see both it and the Milky Way begin to be distorted by their mutual gravitational attraction. What we will not see, as we pointed out earlier, is some sort of wholesale collision between stars. The galaxies are just too empty for that to happen. The "collision" will actually proceed like this: Andromeda will pass through the Milky Way, with gravity distorting both galaxies. When Andromeda has passed through the Milky Way, gravity will slow it down and pull it back for another passthrough. This back and forth will go on for a while until everything settles down and the composite structure (which has been given the uninspiring name of "Milkomeda") becomes just one more elliptical galaxy.

While this is going on, our observer will start to notice something else in the sky. The stars will start to go out: first the big, bright ones—stars that are profligate in their use of hydrogen and burn out their spectacular lives in a few tens of millions of years—followed by smaller stars that live more sedate lives and last longer. Eventually, however, even the most abstemious stars—small, dim objects in the sky—use up their fuel and wink out of sight. With no unburned hydrogen left to fuel them, stars become things seen in the rear-view mirror of the universe. If it matters at this point, these small stars will disappear in about a trillion years.

At the same time, the accelerating Hubble expansion will begin to carry distant galaxies away from us, until, one by one, they pass the point where the distance between them and our observer becomes so great that the light they emit will never reach us. Like the stars, the galaxies will start to wink out of existence.

Over uncounted eons the supermassive black holes now at the centers of thriving galaxies will wander through the darkening universe, picking up the crumbs left over from the death of stars. If two supermassives get near each other, they may start into the death spiral described in chapter 11, as may any stellar black holes still left. Eventually we can imagine a dark universe in which the only thing left of our familiar world is some random radiation and wandering supermassive black holes.

Then what?

This brings up the question of the lifetime of supermassives. In chapter 15 we saw that black holes slowly lose their mass through the process of Hawking radiation, essentially evaporating away like a puddle on a sunny day. The general rule is that the more massive a black hole is, the slower the evaporation process, so that supermassive black holes have incredibly long lifetimes, on the order of $10^{100}$ years. The lifetime of the universe itself is "only" some $10^{10}$ years, which means you would have to wait for ninety more zeroes to accumulate as you watch a supermassive dwindle away. For all intents, we can simply agree that the black hole period in the life of the universe will be long and cold and dark.

There is, however, one caveat we have to make at this point. Should a black hole encounter a chunk of matter—an asteroid, perhaps, or a wandering white dwarf—that matter will be taken inside the Schwarzschild radius of the black hole. This process will increase the mass of the black hole and in effect reset the black hole's clock. Like fictional vampires, in other words, black holes can rejuvenate themselves and extend their lifetime by absorbing fresh material.

Having made that point, we should recognize that toward the end of the black hole era there isn't a whole lot of material left to be absorbed. We doubt, for example, that this absorption process will add even one more zero to our time count.

All of which brings us to the final event in the black hole era. According to some current theories, black holes do not go quietly into the darkness. As they get smaller and smaller, the rate at which they are emitting energy climbs. Calculations suggest that in the final tenth of a second of a black hole's life, it produces an explosion that is the equivalent of a million nuclear fusion bombs going off. This is more than the energy that would be released by all of the nuclear weapons in all of the world's arsenals. This is a huge explosion in human terms, but not by astronomical standards—supernovae trillions of times more energetic have been observed. The explosions of dying black holes will be the only events our hypothetical observer will see throughout these last long eons of existence.

It seems that no matter which possible ending of the universe we choose, we wind up at the same place. Big Rip, Big Crunch, or Flat Universe, we end up in a dark universe populated by isolated particles forever looking for partners with which to interact but never finding them. On the human scale, all our efforts, all our dreams, and all our accomplishments will vanish as the universe descends into that last dark place.

# EPILOGUE

————

## Every atom belonging to me as good belongs to you.

*Walt Whitman*, Song of Myself

Twenty years have passed since I, Shobita, gave up driving trucks down mountains. I sit in my office waiting. I'm waiting for the James Webb Space Telescope (JWST) to point its eighteen hexagonal mirrors, each plated in gold, at a very tiny galaxy 100,000 times smaller than our own. Inside this dwarf galaxy, we think, resides a black hole that resembles the original seeds from which supermassive black holes formed in the early universe. We now know that supermassive black holes existed very early in the universe, when the universe was only a few hundred billion years old in its 13.5-billion-year history. They must have formed very early in the universe, sometime soon after the Big Bang. We cannot observe them forming directly, since the light that they would emit as they feed on primordial gas is beyond the reach of current telescopes, even JWST. We think this tiny galaxy is a fossil remnant of infant galaxies that will give us a clue to the very first supermassive black holes that were formed in the primordial universe. I'm keeping my fingers crossed that inside this tiny galaxy resides that seed of a supermassive black hole. I am hoping that we can measure the mass of this seed, and I am hoping that it will help us understand how the monstrous black holes that lie in the hearts of virtually all massive galaxies formed. Inside may reside an intermediate mass black hole, maybe only a thousand times the mass of the Sun. If so, it will be the first time a black hole in this mass range has been identified.

As we saw in chapter 19, the origins of supermassive black holes are one of the biggest mysteries we are trying to solve. We saw in chapter 4 that black holes form from the death of massive stars, and there is irrefutable evidence that those black holes exist in our galaxy. Yet their masses are only on average about ten times the mass of the Sun. We saw in chapter 5 that the spectacular source Cygnus X-1 is only twenty-one times the mass of the Sun. Even bigger black holes have been found by LIGO, with a record so far of 142 times the mass of the Sun, as we saw in chapter 11. But then there are the supermassives, the smallest of which is about ten thousand times the mass of the Sun, but most of which exceed a million times the mass of the Sun. There is a huge gap in mass of black holes known in the universe. Where are they? It's like we landed on a planet and saw babies and old people but nothing in between. Do the babies evolve into the larger older adults, or are they completely separate beings? The formation and evolution of these elusive beasts remains a mystery. We saw in chapter 8 that one of the common ways to detect black holes is to measure the effect they have on the motion of the surrounding gas and stars. But when the black hole mass is in this intermediate range, the gravitational influence it has on the host galaxy is difficult to detect. And if it is accreting, it won't shine as brightly as the quasars, since smaller black holes won't be as bright. That's why we need JWST. We need a very sensitive telescope that can peer into the interiors of the smallest galaxies, where we think this population of intermediate mass black holes are hiding. We picked a galaxy that has some signs that it might host a feeding intermediate mass black hole. If it does, JWST should see it and should be able to tell us how it feeds and what it does to its host galaxy. All of this will help us discriminate between various models on supermassive black hole seed formation in the early universe.

JWST is revolutionary. When I was driving that truck down the mountain, it was simply a sketch of an amazing idea on a napkin. On the day of the launch, John Mather, the senior project scientist, remarked with amazement at all that it took to take that sketch and turn it into 6.6 tons of flight hardware. It took more than ten thousand dedicated and persistent scientists to make this day happen. When it was conceived, I was a college student. At that time, most of the technology that now sits around one million miles from Earth didn't exist. There were no large aperture telescopes launched in space. JWST needs to peer into the early universe, to see the very first galaxies that were formed. To do this, it needed to be large, large enough to have the collecting area to see the light emitted from the faintest galaxies only a few hundred million years after the Big Bang. The optics had to be lightweight in order to be able to launch them into space, and they had to be stable at cryogenic temperatures and stiff so they hold their shape as they cool down. That's not easy. The Palomar five-meter telescope in California weighs fourteen tons. That's almost the weight of three African elephants. The mirrors were made of beryllium, which is extremely strong for its weight. These lightweight beryllium mirrors had to make their way through eleven different places around the United States to complete their manufacturing. They came to life deep in the depths of beryllium mines in Utah and then moved across the country for processing and polishing. This is not easy. Beryllium is a carcinogen, highly toxic to machine or even touch. The deployable structures, needed to allow the mirror to fit in the rocket fairing, had never been tested. The detectors didn't exist. The cryogenic actuators didn't exist. And the instruments that are now collecting data use technologies that were just a dream a generation before they were deployed.

We have come a long way from observing at the top of a 14,000-foot (4,267 m)-tall mountain. Infrared astronomy is not hard only because infrared light cannot penetrate Earth's atmosphere. The atmosphere itself emits infrared radiation, making it difficult to "see" the infrared light from astronomical sources when the detectors are swamped with light emitted by the sky, the telescope, and the observatory around it. It's like trying to see the stars with your eyes in broad daylight. But a million miles from Earth, the infrared background radiation is more than a million times fainter than it is from Mauna Kea, and the telescope and instruments can get really cold to reduce the infrared light that they themselves emit. JWST needed to be that far away, and being that far away and completely unserviceable was frightening. The community held its breath on that day and in the months that followed. JWST had 344 single point failures when it left Earth. The primary mirror itself had 178 release mechanisms. Everything had to function perfectly in unison. Nothing could go wrong. Every astronomer on the planet waited in silence and solidarity. Hearts were pounding, and a keen awareness of our collective humanity was felt throughout our community. At 7:20 a.m. EST on December 25, 2021, came a collective exhalation by every astronomer across the globe as we watched in awe as James Webb said goodbye to all of us and began its journey into the unknown.

I am so grateful for the thousands of hands that worked with devotion for so many years, conceiving the idea, designing all the parts of the observatory, polishing the mirrors, working on the detectors, making sure that everything would come together perfectly. It did just that. The launch went smoothly, the deployment went smoothly, the mirror alignment went smoothly. Everything worked perfectly. Within a few months after launch, we all realized with amazement that the observatory performance exceeded expectations.

I'm trying hard to be patient. It hasn't been easy waiting. And I worry. I worry I got the coordinates wrong and this $10 billion observatory will miss our galaxy and point to an empty patch of sky. Our team worked so hard for this day. It took months to write the observing proposal. Many people don't know that astronomers have to first come up with a creative science idea, and then work for months to write a proposal describing their science. The proposal is then reviewed by a panel of dedicated scientific peers who work for free for weeks reviewing hundreds of proposals.

The proposal process is highly competitive. You must argue why your science must be done, and why James Webb is the only facility that can do your science. You must convince your peers that your science will revolutionize our understanding of the cosmos. And then you must demonstrate that technically the observatory can achieve your science goals. What instrument is needed? How much exposure time is needed? What are the instrument modes and visibility windows when the observatory can point to your targets? How much overhead will there be, and is it worth the science you gain? All of these calculations are extremely time-consuming and difficult to do while also trying to put together a compelling science case. I didn't initially realize that because JWST is so sensitive, the exposure time needed to observe our tiny galaxy was only a few minutes, but the slew time to go from one target to another is more than an hour. This is because JWST is a floppy telescope. It takes time to align the mirrors. The overheads to observe any target are very large.

So every target must be well justified, and the number you observe must be critical to the science goals of your project. You must argue why the amount of JWST time you request is crucial. Last fall, the Space Telescope Science Institute, flight operation center for Webb, had received 1,931 unique proposals, submitted by 6,291 unique investigators and 24,031 coinvestigators from across the world, representing fifty-six countries, forty-seven states, and two U.S. territories. More than 48,000 hours of observing time were requested, and only about 5,000 hours were granted. We were over-the-moon excited when we got time.

I have extreme gratitude for every photon we manage to collect. I must have checked the coordinates of our targets more than twenty times. I pay attention to all the news. Scheduling is extremely complicated on Webb. We don't actually know when our data will be collected until perhaps a few weeks before it is officially scheduled. Many factors go into scheduling to ensure maximum efficiency of telescope time, and any unforeseen circumstances may affect scheduling. What is happening with the micrometeor hits? There have been more than sixty tiny hits, but the mirror is still performing beyond expectations. As a precaution, JWST now tries to minimize the time it spends looking in the direction in which it is orbiting. This is because when it does that, the micrometeoroids will be moving at higher velocities, potentially causing damage to the mirror. As a result, the telescope's operations have been adjusted to avoid increasing the chance of hits by high velocity micrometeoroids. I also listen to the news about the Artemis launch, because I know that will affect JWST operations. Webb communicates with the ground using the Deep Space Network. The DSN has three sites around the world: one in Goldstone, California; one in Canberra, Australia; and one in Madrid, Spain. This allows us to communicate with Webb at any time of day as Earth rotates. But I didn't realize that the DSN has to be shared with forty different missions and that when there is an Artemis launch things get tricky. Webb must download all the data from its recorders, or it will need to sit there holding all it collected, waiting for a chance to connect with Earth. We stay informed about all of this, hoping that the observatory will be well supported and cared for, that it will be performing at maximum efficiency, and that all will go smoothly in the years to come.

Meanwhile, I will write more proposals. With the extremely competitive selection process, the odds are almost 10:1 that any given astronomer will be successful. We need to keep trying, need to keep thinking of innovative ideas. We need always to be working very hard. I also write more proposals because I constantly worry about my students. I worry that they need to have cutting-edge projects to help build their careers, and I need to pay them, and the only way I can do that is through my grants. Every observing proposal I write comes with some funding, not much, but just enough to fund a graduate student or two to support them as they analyze the data. The funding is limited, and universities charge steep and ever-increasing overheads for every research dollar brought in.

So we struggle to keep our groups afloat. One year I wrote twenty-six proposals. I have no control over the due date. I wrote one once while suffering from pneumonia, because if I missed the deadline, I would need to wait an entire year for the next cycle. One of my students is currently unfunded and has no health insurance. I think about this at night and worry about all the different tasks I am managing. I can't afford to fail. Each student is simply a joy to work with. I am deeply grateful that we get to work together and share this precious time trying to analyze this amazing data. I realize that we have a collective call to action, that we are carrying out the work of Galileo and all the curious humans who came before him, and that what we do will help pave the way for those who come after us. I think about how remarkable it is that humanity managed to launch a giant eye a million miles from Earth and that it is currently staring into the origins of the universe.

More than a year has passed since the launch, and JWST has revolutionized our understanding of early galaxies and black holes. To our surprise, many more galaxies are being found very early in the history of the universe, and they are much more massive than originally thought. Somehow stars were formed rapidly and vigorously early in the universe in ways we don't yet understand. We have also found hundreds of faint accreting black holes. Dubbed "little red dots," these tiny monsters are being found up to redshifts when the universe was only 430 million years old. These new active galactic nuclei (AGNs) are far more abundant than we thought based on our understanding of the number density of quasars in the distant universe, and the black holes appear to be much more massive compared to their host galaxies than we find in the local universe. If they are all really confirmed to be accreting supermassive black holes, these discoveries are challenging our understanding of supermassive black hole formation and their connection to galaxy evolution. The theorists' heads are spinning. The observers are hard at work analyzing the data. Slowly we will uncover the conditions of the early universe during the dawn of galaxy and black hole formation.

On June 24, 2023, about a year after science operations began, JWST turned its golden mirror to our little galaxy where we think may hide an intermediate mass black hole. At 10:00 a.m., the giant golden eye turned slowly and stared for eight hundred seconds at the galaxy. We waited with anticipation, all of us praying that we did not make a mistake. I must have checked the coordinates of our galaxy ten times in the week before our observation, but still I had nightmares that I made a mistake. I imagined receiving data of nothing, simply blank sky. It has happened to other unfortunate investigators, and the time is irrecoverable. Please let it not be us.

We waited with anticipation. I could not sleep that night. The next morning, I received an email with a link to the data file. I hesitated a moment, and then I clicked the link.

A large team of scientists and software developers worked tirelessly to do the preliminary processing of the data so that the products astronomers receive can be analyzed. Because the observatory is new and functioning beyond expectations, a lot of the work that must be done to process the data takes time. But my initial click to the data repository was mind-blowing. I did not type in the wrong coordinates. The galaxy, an extremely faint speck on the sky as seen through ground-based telescopes, is smack in the middle of the image, and it is spectacular. With JWST's spectacular spatial resolution, we can see intricate details in the center of this compact galaxy. We can see young stars forming, and we can see a bright central source. Maybe it's an intermediate mass black hole! We examine the spectra coming from the central bright source and we cannot believe our eyes. There are more than two hundred spectral lines we can see over only a narrow range in wavelength. On my days at Mauna Kea, we could see only three or four lines in the same spectral region in galaxies a million times brighter. But in just a few minutes of staring at this smudge in the sky, JWST uncovered a treasure trove of spectral signatures. Encoded in these spectral features is the information we need to see whether they are produced by an intermediate mass black hole or by young massive stars that are vigorously forming in the center of this tiny galaxy, one million times less massive than our Milky Way. It will take months of hard work by our team to properly analyze the data, to carry out detailed simulations, so that we can understand what is producing the spectral signatures that illuminate our computer screen.

I stare in awe at the data, and I am profoundly moved by what I see. The spectrum contains a multitude of spectral signatures from hydrogen, the most abundant element in the universe. There are very few spectral signatures from any other atom. This is because this galaxy did not evolve much since the beginning of the universe. In fact, it is very much like the more distant galaxies that JWST is uncovering only a few hundred million years after the Big Bang. This galaxy is nearby, though, so we can study it in exquisite detail. It's a perfect laboratory to study the physics of how the first stars and supermassive black holes form. Because the galaxy is a primordial analogue, there have been no previous generations of stars that have lived and died and created the heavy elements that constitute you and me. I stare with wonder as I realize that we are staring into our very origins, deep in the interiors of the first generation of stars. Every atom whose spectral signature has landed on the telescope and has been downloaded to us went into creating the atoms from which my eyes are made that allow me to see this spectrum. And as the hydrogen gets churned in the bellies of these stars to make the heavier elements, I am aware that they eventually went into assembling all of us—all of us who somehow worked together to put a giant golden eye one million miles above our heads to look back at ourselves.

# INDEX